Springer Theses

Recognizing Outstanding Ph.D. Research

For further volumes:
http://www.springer.com/series/8790

Aims and Scope

The series "Springer Theses" brings together a selection of the very best Ph.D. theses from around the world and across the physical sciences. Nominated and endorsed by two recognized specialists, each published volume has been selected for its scientific excellence and the high impact of its contents for the pertinent field of research. For greater accessibility to non-specialists, the published versions include an extended introduction, as well as a foreword by the student's supervisor explaining the special relevance of the work for the field. As a whole, the series will provide a valuable resource both for newcomers to the research fields described, and for other scientists seeking detailed background information on special questions. Finally, it provides an accredited documentation of the valuable contributions made by today's younger generation of scientists.

Theses are accepted into the series by invited nomination only and must fulfill all of the following criteria

- They must be written in good English.
- The topic should fall within the confines of Chemistry, Physics, Earth Sciences, Engineering and related interdisciplinary fields such as Materials, Nanoscience, Chemical Engineering, Complex Systems and Biophysics.
- The work reported in the thesis must represent a significant scientific advance.
- If the thesis includes previously published material, permission to reproduce this must be gained from the respective copyright holder.
- They must have been examined and passed during the 12 months prior to nomination.
- Each thesis should include a foreword by the supervisor outlining the significance of its content.
- The theses should have a clearly defined structure including an introduction accessible to scientists not expert in that particular field.

Eva Maria Huber

Structural and Functional Characterization of the Immunoproteasome

Doctoral Thesis accepted by
the Technical University Munich, Germany

Author
Dr. Eva Maria Huber
Chair of Biochemistry
Technische Universität München
Garching
Germany

Supervisor
Prof. Dr. Michael Groll
Chair of Biochemistry
Technische Universität München
Garching
Germany

ISSN 2190-5053 ISSN 2190-5061 (electronic)
ISBN 978-3-319-37810-7 ISBN 978-3-319-01556-9 (eBook)
DOI 10.1007/978-3-319-01556-9
Springer Cham Heidelberg New York Dordrecht London

Printed on acid-free paper

Springer is part of Springer Science+Business Media (www.springer.com)

Parts of this thesis have been published in the following journal articles:

Immuno- and constitutive proteasome crystal structures reveal differences in substrate and inhibitor specificity
Huber, E. M.*, Basler, M.*, Schwab, R.*, Heinemeyer, W., Kirk, C. J., Groettrup, M., Groll, M. (2012), Cell, *148*, 727–738.

The 19S cap puzzle: A new jigsaw piece
Huber, E. M. and Groll, M. (2012). Structure, *20,* 387–388.

Kristallstruktur eines molekularen Schredders
Huber, E. M. and Groll, M. (2012). GIT Labor-Fachzeitschrift, *5*, 363–365.

Inhibitors for the immuno- and constitutive proteasome: current and future trends in drug development.
Huber, E. M. and Groll, M. (2012). Angew. Chem. Int. Ed. Engl., *51,* 8708–8720.

*These authors contributed equally

The work in this thesis was conducted from September 2009 to January 2013 under the supervision of Prof. Dr. Michael Groll, Chair of Biochemistry, TUM.

To my family

Supervisor's Foreword

Curing diseases is the primary long-term goal of contemporary scientific research. However, developing new drugs and bringing them to application demands for enormous efforts and staying power of both academia and pharmaceutical industry. Often the structural analysis of a target protein and its understanding at the atomic level form the foundation of such a long-lasting process. This proved true also for the 20S proteasome core particle (CP), a protease of 720 kDa and 28 single subunits.

In 1995, the first X-ray structure of a 20S proteasome, namely that of the archaeon *Thermoplasma acidophilum* has been elucidated by Löwe and coworkers [1]. Only, 2 years later, the structure of the proteasome from baker's yeast—the first eukaryotic proteasome—was solved [2]. This milestone in proteasome research stimulated the development of the multitude of inhibitory compounds that is known today. Many of these drugs have been structurally analyzed in complex with the yeast 20S proteasome and the obtained X-ray data served as intermediate and validation steps in the drug design development process. Up to now, two proteasome inhibitors have made their way from bench to bedside: Velcade® and Kyprolis®. Since the proteasome is essential for many cellular processes including cell cycle progression, both compounds are applied to patients suffering from blood cancer. However, recently, a novel therapeutic application of proteasome inhibition has been discovered. The compound ONX 0914 (formerly PR-957)—identified in high-throughput screenings—was shown to be of therapeutic benefit in animal models of autoimmune diseases such as rheumatoid arthritis and lupus erythematodes [3, 4]. Remarkably, despite its pronounced structural similarity to other proteasome inhibitors, ONX 0914 selectively targets only the 20S immunoproteasome, a specialized version of the proteasome known in vertebrates.

The immunoproteasome selectivity of ONX 0914 fascinated me as a chemist and formed the starting point for the Ph.D. thesis of Dr. Huber. Her work culminated in the first X-ray structure of an immunoproteasome. Finally, apo and ligand complex structures of the immunoproteasome with ONX 0914 provided an explanation for its selectivity. The structural results described herein represent a valuable contribution for modeling and designing novel proteasome-type selective and subunit-specific inhibitory compounds. Apart from the immunoproteasome structure described herein, Dr. Huber conducted yeast mutagenesis experiments which aimed at imitating all three 20S proteasomes types of vertebrates in yeast: the constitutive proteasome, the immunoproteasome and the thymoproteasome. Together, the topic and the results of this thesis will definitely inspire further efforts in academic research as well as in medicinal chemistry.

Munich, June 2013 Prof. Michael Groll

References

1. J. Löwe, D. Stock, B. Jap, P. Zwickl, W. Baumeister, R. Huber, Crystal structure of the 20S proteasome from the archaeon *T. acidophilum* at 3.4 Å resolution. Science **268**, 533–539 (1995)
2. M. Groll, L. Ditzel, J. Löwe, D. Stock, M. Bochtler, H.D. Bartunik, R. Huber, Structure of 20S proteasome from yeast at 2.4 Å resolution. Nature **386**, 463–471 (1997)
3. T. Muchamuel, M. Basler, M.A. Aujay, E. Suzuki, K.W. Kalim, C. Lauer, C. Sylvain, E.R. Ring, J. Shields, J. Jiang, P. Shwonek, F. Parlati, S.D. Demo, M.K. Bennett, C.J. Kirk, M. Groettrup, A selective inhibitor of the immunoproteasome subunit LMP7 blocks cytokine production and attenuates progression of experimental arthritis. Nat. Med. **15**, 781–787 (2009)
4. H.T. Ichikawa, T. Conley, T. Muchamuel, J. Jiang, S. Lee, T. Owen, J. Barnard, S. Nevarez, B.I. Goldman, C.J. Kirk, R.J. Looney, J.H. Anolik. Novel proteasome inhibitors have a beneficial effect in murine lupus via the dual inhibition of type i interferon and autoantibody secreting cells. Arthritis Rheum. **64**, 493–503 (2011)

Acknowledgments

First of all I would like to thank my Supervisor Prof. Dr. Michael Groll for providing me the possibility to work in his excellent team and for entrusting me the immunoproteasome project. I am very grateful for his daily interest in my work, his deep confidence in me, his continuous support, and substantial promotion. His enormous enthusiasm and encouragement as well as his advices and ideas were of great help not only for my scientific results but, even more important, also for my personality.

Special thanks go to my collaborators from the University of Constance Prof. Dr. Marcus Groettrup, Dr. Michael Basler, and Ricarda Schwab. Without their contribution—the purification of the murine proteasomes—this work would not have been possible.

I strongly thank PD Dr. Wolfgang Heinemeyer for sharing his outstanding knowledge and experience in yeast genetics with me, for helping me creating the numerous mutants and for all the new techniques he taught me.

I am very grateful to Dr. Melissa Gräwert who patiently introduced me to the theory and the practice of crystallography during the first year of my Ph.D.

Moreover, I want to acknowledge Richard Feicht for all the yeast proteasome purifications and the amazing crystals he produced.

I am also indebted to all the students who joined me in the lab. In particular, I thank Silvia Domcke for her excellent work.

Last but not least, I want to thank all members of the Groll group, especially my Ph.D. colleagues for the nice and relaxed working atmosphere, one of the things I appreciated and enjoyed most during the last 3 years. I am grateful for all the funny moments inside and outside the lab, especially during the synchrotron trips and for all the group activities, we did together. Especially, I thank Ute and Astrid for managing many of my problems with forms and orders.

Finally, I thank my family for having enabled my studies, for their support, their confidence in me, and their interest in my work.

Contents

Abbreviations

Å	Ångström
aa	Amino acid
Ac	Acetate/acetyl
AMC	7-Amino-4-methylcoumarin
AMP	Adenosine monophosphate
Amp	Ampicillin
APS	Ammonium persulfate
ATP	Adenosine triphosphate
B. taurus	*Bos taurus*
bp	Base pairs
Braap	Branched chain amino acid preferring
BSA	Bovine serum albumin
°C	Degree Celsius
Cbz	Carboxybenzyl
cCP	20S Constitutive proteasome
CD8	Cluster of differentiation 8
ChTL	Chymotrypsin-like
CL	Caspase-like
CM	Complete medium
CP	Core particle, 20S proteasome
cTEC	Cortical thymic epithelial cells
CTL	Cytotoxic T lymphocyte
2D	Two-dimensional
3D	Three-dimensional
Da	Dalton
ddH_2O	Double distilled water
DMSO	Dimethylsulfoxid
DNA	Deoxyribonucleic acid
dNTP	Deoxyribonucleotide triphosphate

DTT	Dithiothreitol
E. coli	*Escherichia coli*
EDTA	Ethylenediaminetetraacetic acid
ER	Endoplasmic reticulum
EtOH	Ethanol
FDA	U. S. food and drug administration
5-FOA	5-fluoroorotic acid
HCl	Hydrochloric acid
Hsp	Heat shock protein
IC_{50}	Half maximal inhibitory concentration
iCP	20S Immunoproteasome
IFN	Interferon
IL	Interleukin
K	Kelvin
kbp	Kilo base pairs
kDa	Kilo Dalton
LB	Luria Bertani
LCMV-WE	Lymphocytic choriomeningitis virus strain WE
LiAc	Lithium acetate
mA	Milliampere
MDa	Mega Dalton
MES	2-(*N*-morpholino) ethanesulfonic acid
MHC	Major histocompatibility complex class
MPD	2-Methyl-2,4-Pentanediol
NCS	Non-crystallographic symmetry
NEPHGE	Non-equilibrium pH gradient gel electrophoresis
NF-κB	Nuclear factor-κB
Ntn	N-terminal nucleophile
OD	Optical density
PAGE	Polyacrylamide gel electrophoresis
PCR	Polymerase chain reaction
PDB	Protein Data Bank
PEG	Polyethylene glycol
pNA	Para-nitroaniline
R_{free}	Free R-factor
r.m.s.d.	Root-mean-square deviation
rpm	Rounds per minute
R_{work}	Crystallographic R-factor
S	Svedberg
S.cerevisiae	*Saccharomyces cerevisiae*
SDS	Sodiumdodecylsulfate
SLS	Swiss Light Source
Snaap	Small neutral amino acid preferring
SOC	Super optimal broth with catabolite repression
Suc	Succinyl

T. acidophilum	*Thermoplasma acidophilum*
TAE	Tris-Acetate-EDTA
TAP	Transporter associated with antigen processing
tCP	20S thymoproteasome
TCR	T cell receptor
TE	Tris-EDTA
TEMED	*N,N,N',N'*-tetramethylethylenediamine
TL	Trypsin-like
TLS	Translation, Libration, Screw
T_m	Melting temperature
TNF	Tumour necrosis factor
Tris	Tris (hydroxymethyl-) aminomethane
Tris-HCl	Tris (hydroxymethyl-) aminomethane hydrochloride
U	Unit
UV	Ultraviolet
V	Volt
VIS	Visible
v/v	Volume per volume
wt	Wildtype
w/v	Weight per volume
yCP	Yeast 20S proteasome
YPD	Yeast extract peptone dextrose

Chapter 1
Introduction

1.1 Intracellular Protein Degradation and Antigen Presentation

All proteins undergo a life cycle, starting from their synthesis at ribosomes and ending with their degradation to peptides and single amino acids. Thereby, the building blocks for the *de novo* synthesis of polypeptides are recycled and in addition important cellular functions are controlled. These include protein homoeostasis, cell proliferation, signal transduction and antigen production [1]. The lysosomal and the non-lysosomal protein degradation pathways process the majority of intracellular self and nonself proteins. In lysosomes proteases termed cathepsins unselectively degrade proteins. The resulting peptide fragments can be loaded on major histocompatibility complex class (MHC) II receptors and presented on the cell surface to immune cells [2]. In contrast, hydrolysis of more than 90 % of all cytosolic proteins is carried out by the 26S proteasome—the central player of the non-lysosomal protein degradation pathway. The 26S proteasome, a multicatalytic ATP-dependent protease located in the cytosol and the nucleus, selectively cuts polyubiquitylated proteins to peptides of diverse lengths [3, 4]. While most of these peptides are further decomposed to single amino acids, a fraction escapes this fate and serves as antigens for the immune system in vertebrates. Prior or after N-terminal trimming to 8–11 amino acids by aminopeptidases [5, 6] putative epitopes transit into the endoplasmic reticulum (ER) by the transporter associated with antigen processing (TAP), a member of the ATP-binding cassette transporter family [7]. In the ER peptides can associate with the binding cleft of nascent MHC I receptors. The peptide's affinity for the MHC I complex largely depends on its C-terminal anchor residue [8, 9], which is determined by the cleavage specificities of the proteasome. Only stable receptor:ligand complexes adopt a mature structure and are transported in vesicles to the cell membrane for their exposure to the extracellular environment. Surveying effector cells of the immune system scan the peptide cargos of both MHC I and II receptor proteins for their origin and cells that carry bacterial or viral peptides on their cell surface are identified

E. M. Huber, *Structural and Functional Characterization of the Immunoproteasome*,
Springer Theses, DOI: 10.1007/978-3-319-01556-9_1,

and eliminated by cytotoxic T lymphocytes to prevent an infection [10, 11]. Hence, by shaping the antigenic pool of peptides the proteasome constitutes a key component of the adaptive immune system of vertebrates.

1.2 The Ubiquitin–Proteasome System

1.2.1 Polyubiquitylation: A Signal for Degradation

Posttranslational modifications such as ubiquitylation control the subcellular localization, signal transduction, enzymatic activity and stability of eukaryotic proteins. Ubiquitin itself is a highly conserved polypeptide of 8.5 kDa and 76 amino acids and its covalent linkage to proteins is catalysed by the sequential action of three classes of enzymes, termed ubiquitin-activating enzyme (E1), ubiquitin-conjugating enzyme (E2) and ubiquitin ligase (E3) [4]. The E1 enzyme activates the C-terminal glycine residue of ubiquitin in an ATP-dependent manner under the release of pyrophosphate. Subsequent attachment of the adenylated ubiquitin to the active site Cys residue of the E1 enzyme yields a thioester and an AMP molecule. Next, ubiquitin is conjugated to the catalytic Cys of an ubiquitin-conjugating enzyme (E2) and finally, an E3 ligase creates an isopeptide bond between the ubiquitin's C-terminus and the ε-amino group of a lysine side chain in the substrate protein (Fig. 1.1) [4]. Selectivity of the ubiquitylation reaction is ensured by a myriad of E3 ligases that binds to recognition sites in their target proteins either directly or via adaptor proteins. Commonly, proteins that carry a chain of at least four ubiquitin molecules connected via Lys48 are targeted for degradation by the 26S proteasome [12].

1.2.2 Subunit Composition and Architecture of Proteasomes

The 26S proteasome is a molecular machine of approximately 2.5 MDa, consisting of a cylindrical 20S core particle (CP) and two 19S regulatory complexes associated on both ends. The 19S cap binds polyubiquitylated substrates, cleaves off their ubiquitin molecules, unfolds and finally translocates client proteins into the proteolytic 20S chamber [14]. In contrast to the 19S complex whose atomic structure has not yet been resolved, the core of the proteasome has been extensively characterized at the molecular level, leading to the following current understanding of its evolution and functioning.

The barrel-shaped CP is present in all three kingdoms of life, bacteria, archaea and eukaryotes. Bacteria except for actinomycetes harbour CP-like complexes termed heat-shock locus (Hsl) V that consist of two homohexameric rings of a

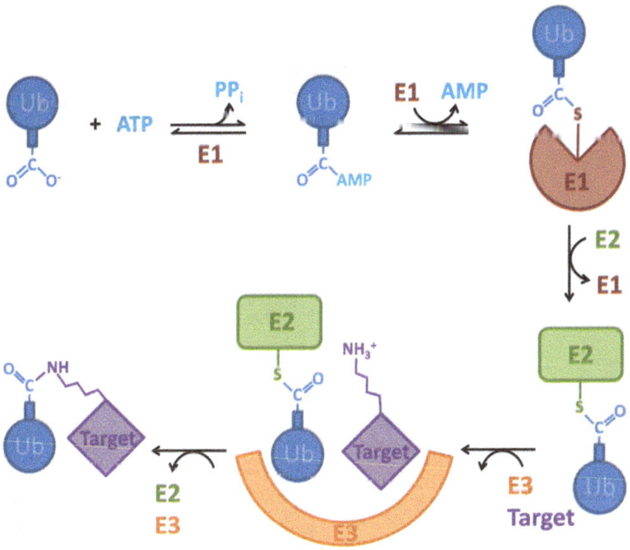

Fig. 1.1 Ubiquitin activation and linkage to substrate proteins. Adenylation of ubiquitin by an El enzyme is followed by its transfer to an E2 ubiquitin-conjugating enzyme. An E3 ligase covalently attaches ubiquitin to a lysine residue in the target protein. Adapted from Berg et al. [13]

proteolytically active subunit [15]. The CP of actinomycetes, archaea and eukaryotes adopts a more complex quarternary structure. The X-ray structures of the CP from the archaeon *Thermoplasma acidophilum* [16] and the eukaryote *S. cerevisiae* [17] proved that the protease is a cylinder of 148 Å in length and 113 Å in width with a molecular mass of approximately 720 kDa and that it consists of two different types of subunits termed α and β. Both monomers are assembled in four homoheptameric rings that are stacked in an αββα stoichiometry around a central pore. Due to its two-fold rotational symmetry the CP comprises two identical halves (Fig. 1.2) [16, 17]. Archaeal 20S proteasomes incorporate only one type of α and one type of β subunit [16]. By contrast, eukaryotic proteasomes contain seven different α (1–7) and seven different β (1–7) subunits (Fig. 1.2) [17], which occupy unique positions within the CP. All known α and β subunits share the same tertiary structure of two antiparallel five-stranded β sheets being flanked by α helices and probably evolved from a common ancestor (Fig. 1.2a) [16, 18].

Several functions have been attributed to the α subunits. Their intrinsic ability to form a heptameric ring is pivotal for the assembly of the β subunits and the formation of the CP. Furthermore, so-called shuttling sequences in the α subunits enable the import of the CP into the nucleus and its re-export [23]. The α subunits also form the entry gate to the interior of the CP [24, 25]. In particular, the N-terminal extensions of the subunits α2, α3 and α4 close the barrel-shaped CP on

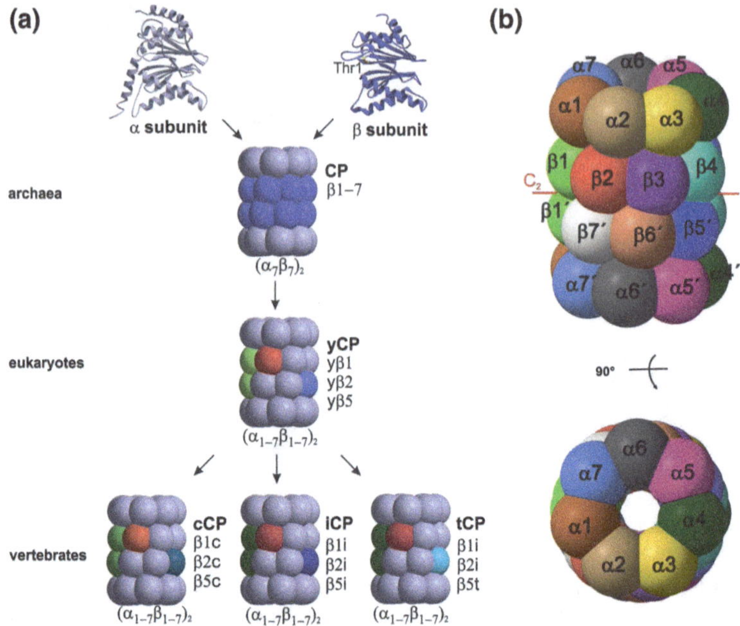

Fig. 1.2 Evolution of 20S proteasomes. (**a**) Archaeal CPs are assembled of two different types of monomers, namely inactive α (*grey*) and proteolytically active β subunits (*blue*). Both presumably originated from a common but unknown precursor protein [18]. Eukaryotes such as *S. cerevisiae* encode seven different α and seven different β subunits, but only the β subunits β1 (*green*), β2 (*red*) and β5 (*blue*) are enzymatically active [17]. The inactive α and β subunits are coloured in *grey*. In vertebrates three classes of 20S proteasomes evolved. The constitutive proteasome (cCP), the immunoproteasome (iCP) and the thymoproteasome (tCP) have distinct sets of catalytic β subunits and thus, are implemented in different biological pathways [17, 19–21]. (**b**) *Side* and *top view* of the eukaryotic CP. The C_2 symmetry of the proteasome particle is indicated as a *red* rod. Adapted from Huber and Groll [22]

both ends and thereby abolish uncontrolled degradation of intracellular proteins [16, 17]. Docking of regulatory and adapter complexes such as the 11S (PA28), 19S and PA200 (Blm10) particles onto the α ring rearranges the N-termini of the α subunits and enables access to the proteolytically active sites that are sequestered in the β rings [26]. The β subunits are synthesized as inactive precursor proteins with N-terminal propeptides of up to 75 amino acids. These propeptides are involved in CP assembly assisted by the proteasome maturation factor Ump1 (ubiquitin-mediated proteolysis 1) [27] and protect the catalytic active sites from inactivating Nα acetylation prior to proteasome maturation [28]. In the final step of proteasome assembly the propeptides are removed by intramolecular autolysis, thereby exposing the proteolytically active Thr1 residues and yielding a proteasome particle that can cut polypeptides down to size [16, 29]. Remarkably, in eukaryotes only three out of the seven different β subunits, namely β1, β2 and β5, are proteolytically active as N-terminal nucleophile (Ntn) threonine hydrolases [17].

Whereas primitive eukaryotes like baker's yeast bear only one type of CP [17], vertebrates express three classes of CPs [21, 30]: the constitutive proteasome (cCP), the immunoproteasome (iCP) and the thymoproteasome (tCP). The cCP comprises the catalytic active constitutive (c) subunits β1c (Y, *PSMB6*), β2c (Z, *PSMB7*) and β5c (X, *PSMB5*) and represents the prevailing proteasome species in cells of non-haematopoietic origin. In contrast, in immune cells such as lymphocytes and monocytes the iCP is predominant [30] (Fig. 1.2a). Nonetheless, expression of the iCP can also be induced in non-immune tissues by proinflammatory cytokines such as interferon (IFN)-γ and tumour necrosis factor (TNF)-α [20, 31]. As the three proteolytically active immuno (i) subunits β1i (LMP2, low molecular weight protein 2, *PSMB9*), β2i (MECL1, multicatalytic endopeptidase complex-like-1, *PSMB10*) and β5i (LMP7, *PSMB8*) are preferentially assembled into CPs, mainly iCPs are formed *de novo* upon cytokine release [20, 32]. Apart from the i subunits IFN-γ also triggers the expression of the 11S adaptor complex PA28αβ [33]. This particle stimulates proteasome activity and is essential for the generation of certain antigenic peptides [34]. Furthermore, PA28αβ has been suggested to target iCPs to the TAP in the ER membrane in order to directly translocate the generated peptides into the ER lumen [35]. Besides the cCP and iCP, mixed proteasomes bearing the subunit composition β1c, β2c and β5i or β1i, β2c and β5i were reported to account for 30–50 % of all cellular CPs [36] and even CPs with asymmetric composition of β subunits were described [37].

The third type of CP, the vertebrate-specific thymoproteasome (tCP), is exclusively expressed in cortical thymic epithelial cells (cTECs) and incorporates the i subunits β1i and β2i as well as the exceptional subunit β5t (*PSMB11*) [21] (Fig. 1.2a). The unique expression profile and subunit composition of the tCP has been implicated to play a pivotal role for the development of $CD8^+$ cytotoxic T cells as part of the adaptive immune system.

1.2.3 Proteolysis by 20S Proteasomes

Structural studies along with mutagenesis experiments proved that a sophisticated hydrogen bond network involving the amino acids Thr1, Asp17, Lys33, Ser129, Asp166 and Ser169 is responsible for the nucleophilicity of the γ-hydroxylgroup of the active site Thr1 and thus, for proteolytic activity [16, 17]. A nucleophilic water molecule cluster in the active site induces the proton transfer from Thr1O^{γ} to Thr1N, which functions as proton acceptor. Peptide bond hydrolysis then starts with an attack of the N-terminal Thr1O^{γ} onto the electrophilic carbonyl carbon atom of the scissile peptide bond of substrate proteins (Fig. 1.3). This reaction step creates an acyl-enzyme intermediate and frees the N-terminus of the C-terminal peptide fragment. Addition of a pre-oriented nucleophilic water molecule restores

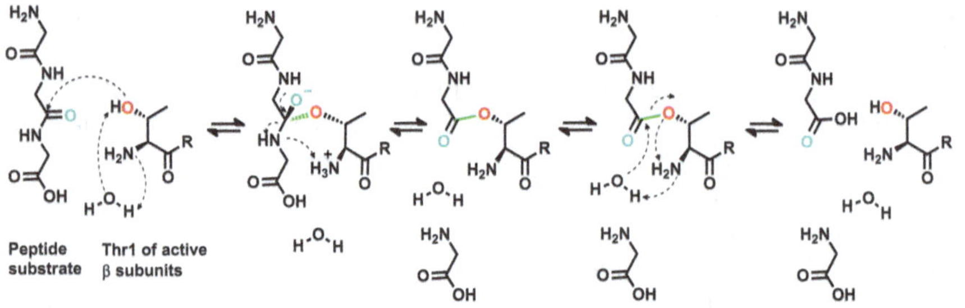

Fig. 1.3 Peptide bond hydrolysis by the proteasome. The nucleophilic Thr1O$^{\gamma}$ (*red*) attacks the electrophilic carbonyl carbon atom of the peptide bond thereby releasing the first cleavage product. In a second step hydrolysis of the formed acyl-enzyme complex (*green* bond) frees the N-terminal cleavage product and restores the catalytic Thr1

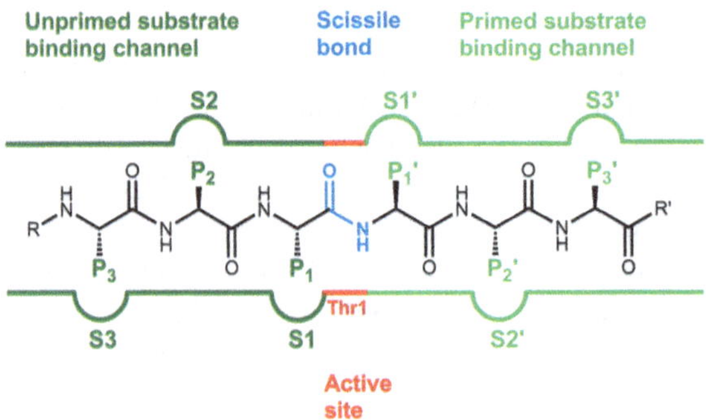

Fig. 1.4 Schematic representation of the substrate binding channel of the proteasomal active sites. The primed pockets (S′; *green*) and unprimed specificity sites (S; *dark green*) bind the ligand's side chains (P′ and P sites). Afterwards the active site Thr1 (*red*) cleaves the scissile peptide bond (*blue*). Adapted from Huber et al. [22, 39]

the catalytic Thr1 and deliberates the C-terminus of the second cleavage product (Fig. 1.3) [38].

The proteasome is classified as an endoprotease, as it harbours primed (S1′, S2′, S3′) and unprimed substrate binding pockets (S1, S2, S3) that are formed by two adjacent subunits [16]. The chemical nature of these sites, which accommodate the side chains (P residues) of client proteins (Fig. 1.4), determines the mean residence time of ligands and thus, cleavage preferences.

In particular the chemical properties of amino acid 45, forming the bottom of the S1 pocket, decide on proteasomal substrate specificities. The positively charged Arg45 of subunit yβ1/β1c favours a caspase-like (CL) activity by accommodating acidic residues in the S1 pocket. However, subunit β1 can also cleave after some hydrophobic amino acids and thereby contributes to the branched chain amino acid preferring (Braap) activity of the proteasome [40, 41]. Even though a trypsin-like (TL) activity has been attributed to subunit yβ2/β2c, it can virtually processes after all kind of residues owing to Gly45. The chymotrypsin-like (ChTL) activity, residing in subunit yβ5/β5c, preferentially hydrolyses proteins C-terminally of hydrophobic amino acids, because its substrate binding channel is lined with apolar residues such as Met45 [17, 41]. In addition, a so-called small neutral amino acid preferring activity (Snaap) has been assigned to the proteasome. For the iCP, previous studies suggested a reduced CL activity, but an enhanced ChTL activity as compared with the cCP/yCP [17, 40, 42, 43]. In contrast, the vertebrate-specific tCP has been demonstrated to exert only minimal ChTL activity [21].

1.2.4 Immune Functions of 20S Proteasomes

CPs shape the pool of antigenic peptides that are presented by MHC I receptors on the cell surface to immune cells. Importantly, all three types of CPs, the cCP, iCP and tCP, are capable of generating antigens [44, 45]. However, their distinct substrate specificities result in different epitope repertoires and physiological implications. Although certain cCP-dependent peptides can trigger immune reactions, the iCP is particularly decisive for this process. Cytokines released during viral infections induce the expression of the iCP in order to enhance antigen presentation and to support clearance of the pathogen. In agreement, mice lacking the iCP bear a significantly altered repertoire of antigens in terms of quantity and quality, implicating that the i subunits enlarge both the abundance and the diversity of MHC I epitopes [46]. Besides, the iCP plays a crucial role in T cell differentiation and the release of proinflammatory cytokines such as interleukin (IL)-23, IL-2 and IFN-γ via a yet unknown nuclear factor-κB (NF-κB)-independent pathway [47]. In addition, proteins that are damaged by cytokine-induced oxidative stress were shown to be efficiently cleared by the iCP [48].

In contrast to the cCP, the expression of the tCP is restricted to the cTECs of the thymus [45]. This specialized organ of the immune system selects T lymphocytes for the optimal interaction strength of their T cell receptor (TCR) with MHC:self-peptide complexes. Immature thymocytes whose TCR tightly binds to MHC receptors loaded with self-antigens are auto-reactive and are eliminated by negative selection, while T cells with weak TCR:MHC interactions are considered as self-tolerant and thus survive (positive selection) [30, 49]. The markedly attenuated ChTL activity of the tCP-specific subunit β5t is assumed to produce low-affinity self-epitopes for the presentation on MHC receptors [45]. Moreover, the peptide pool created by tCPs in cTECs is unique throughout the body and thus, might prevent autoimmunity, as positively selected T cells more seldom cross-react with self-peptides that have been generated by the cCP and iCP outside the cTECs [50]. Remarkably, in β5t-depleted mice the number of CTLs is reduced by 75 %, stressing the importance of the tCP for the maturation of thymocytes [21]. In conclusion, the tCP helps to establish a tolerant T cell repertoire, the prerequisite for a functional adaptive immune system and the iCP promotes immune surveillance and the elimination of pathogens. Hence, both the tCP and the iCP are key players of the adaptive immune system in vertebrates and their malfunctioning positively correlates with the onset of diverse diseases.

1.3 20S Proteasomes: Validated and Emerging Drug Targets

20S proteasomes primarily serve to degrade regulatory or aberrant proteins, but during evolution this function has been exploited for the development of the vertebrate immune system. Thus, the multifaceted functions of CPs make them attractive drug targets for diseases as diverse as cancers and autoimmune disorders.

The CP levels of tumour cells are often upregulated, because their accelerated cell cycle and metabolism require increased turnover rates of proteins. This dependence on CP activity renders neoplastic cells highly susceptible to proteasome inhibition. Among cancers multiple myeloma cells are most sensitive to CP inhibitors [51, 52]. Myeloma cells are derived from plasma cells and their excessive synthesis of immunoglobulins as well as their chromosomal instability leads to many aberrant and misfolded proteins that have to be removed by CPs. Hence, inhibition of the CP causes the accumulation of protein aggregates and finally triggers ER stress as well as the unfolded protein response [53–57]. Furthermore, CP inhibitors block proinflammatory signalling cascades such as NF-κB and the expression of anti-apoptotic target genes [52]. Ultimately, tumour suppressor genes like the cyclin kinase inhibitor $p27^{kip1}$ [58] induce apoptosis in transformed cells, while healthy cells remain unaffected [51]. This difference in susceptibility creates a therapeutic window for proteasome inhibition in haematological cancers.

However, in certain malignancies, including lung, colon and prostate cancers as well as feline primary fibrosarcoma, tumour pathogenesis depends on elevated iCP and cytokine levels [59, 60]. Additionally, Alzheimer's [61] and Huntington's [62] disease as well as amyotrophic lateral sclerosis [63] and inflammatory bowel disease [64] are characterized by increased expression rates of iCPs or single i subunits. Moreover, abnormal levels of i subunits were observed in Sjogren's syndrome [65], inclusion body myositis, myofibrillar myopathy [66, 67], Crohn's disease [68] and dextran sulfate sodium-induced colitis [69]. These findings suggest a therapeutic benefit from blocking solely the iCP, at least for the above mentioned diseases. So far, inhibition of the iCP has proven its effectiveness in a number of autoimmune disorders [47, 70–72], but ambiguous results were obtained with respect to its anti-cancer activity. Although a few malignancies were reported to be sensitive to inhibition of the iCP [60, 73, 74], other studies demonstrate that only simultaneous blockage of several active sites, for instance of β5c and β5i, efficiently causes cell death [75].

In contrast to the cCP and iCP, the therapeutic potential of tCP blockage has not been investigated up to now.

1.4 Types of Proteasome Inhibitors

Drug discovery in the ubiquitin–proteasome system has been considerably facilitated by the crystallographic analysis of the CP from *T. acidophilum* [16], *S. cerevisiae* [17] and *Bos taurus* [19]. Most of the proteasome inhibitors known today harbour an electrophilic head group that reversibly or irreversibly inhibits the catalytic Thr1O$^{\gamma}$. According to their type of pharmacophore covalently acting compounds can be subdivided into seven classes (Fig. 1.5): aldehydes, vinyl sulfones, vinyl amides (syrbactins), boronic acids, α′,β′epoxyketones, α ketoaldehydes (glyoxals) and β lactones. Aldehydes, such as the calpain inhibitors I (Ac-Leu-Leu-norleucinal) and II (Ac-Leu-Leu-methional) were the first inhibitors identified for the proteasome [16, 76]. However, their susceptibility to oxidation and their off-target activity towards serine and cysteine proteases restrict their medical potential. Vinyl sulfones [77] and the naturally occurring syrbactins [78] form an ether bond with Thr1O$^{\gamma}$ in a Michael-type 1,4-addition, but also target cysteine proteases. Furthermore, β lactones are, depending on their R1 substituent, either reversible or irreversible inhibitors of the CP [79] and its most prominent representative marizomib (salinosporamide A; NPI-0052; Nereus Pharmaceuticals, Inc.) is currently in clinical phase I trials for myeloma, lymphoma and leukaemia [80]. Despite their high reactivity and the associated adverse effects, boronic acids such as bortezomib are among the most potent inhibitors of the CP (see also Sect. 1.5). The recently developed second-generation drugs for the proteasome are α′,β′epoxyketone-based compounds. Their pharmacophore is derived from the natural product epoxomicin that irreversibly inhibits the CP in a bivalent reaction involving both Thr1O$^{\gamma}$ and Thr1N (see Sect. 4.3.4) [81, 82]. Due to this mode of action α′,β′epoxyketones are highly selective for Ntn hydrolases such as the 20S proteasome. Employing a reaction mechanism similar to α′,β′epoxyketones, α ketoaldehydes also selectively target CPs but their formation of a cyclic Schiffbase with Thr1O$^{\gamma}$ and Thr1N is reversible [83].

Except for lactones and syrbactins, CP ligands consist of a peptidic backbone of two to four amino acids that is attached to one of the listed electrophilic head groups. These peptide-based compounds were shown to mimic natural protein substrates by forming an antiparallel β sheet in the substrate binding channels of the yCP [84].

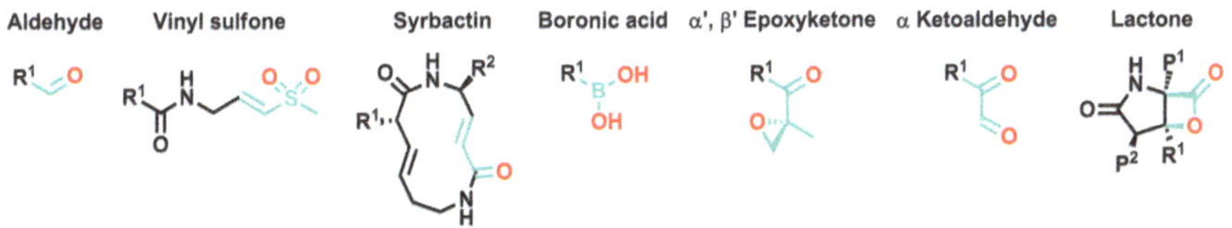

Fig. 1.5 Electrophilic headgroups of covalently acting proteasome inhibitors. The functional groups are shown in *cyan* with their oxygens marked in *red*. R^1 and R^2 designate the variable parts of the compounds. For lactones the P^1 and P^2 side chains, targeting the corresponding S pockets of CPs, are depicted

1.5 Clinically Relevant Proteasome Inhibitors

After years of academic and pharmaceutical research, in 2003, bortezomib (Velcade®, Millenium Pharmaceuticals, Inc.; PS-341) has been granted full approval by the U. S. food and drug administration (FDA) for the treatment of multiple myeloma as well as relapsed or refractory mantle cell lymphoma (Fig. 1.6) [85]. Furthermore, the therapeutic benefit of bortezomib for organ transplantation [86] and solid tumours like non-small cell lung cancer [87] is currently being investigated. With annual sale rates of more than 2 billion dollars bortezomib is a blockbuster drug that substantially elongates the life spans and survival rates of multiple myeloma patients. The dipeptide boronic acid inhibitor potently inhibits the β5c, β5i and β1i active sites of the cCP and iCP with IC_{50} values of 3–8 nM and only in higher concentrations also targets the β1c, β2c and β2i subunits [57]. Boronic acids such as bortezomib form a reversible tetrahedral transition state with Thr1O^{γ} that is stabilized by hydrogen bonds with Thr1N and the oxyanion hole Gly47NH of active proteasome subunits [88]. Although these interactions promote a higher affinity of bortezomib for Ntn hydrolases, bortezomib was shown to also considerably inhibit serine proteases, including cathepsin G, cathepsin A, chymase, dipeptidyl peptidase II and HtrA2/omi, involved in neuronal survival [89]. These off-target activities cause severe neurotoxicity leading to tremor, reduced nerve conduction velocity and nerve degeneration, which affect about 30 % of all patients treated with bortezomib [89]. Additional drawbacks of bortezomib include thrombocytopenia, neutropenia and gastrointestinal disorders as well as its poor bioavailability that necessitates intravenous administration [90].

Moreover, a significant fraction of newly diagnosed patients does not respond to bortezomib and treated patients often relapse [91]. Whereas increased concentrations of the heat shock protein (Hsp) 27 were reported to confer primary resistance [92], the molecular basis for acquired non-responsiveness to bortezomib is still in focus of scientific research efforts. So far, cell culture studies analysing

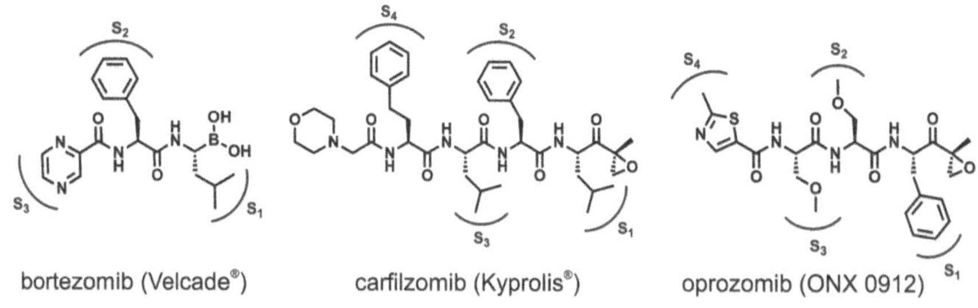

Fig. 1.6 Prominent proteasome inhibitors. The chemical structures of the dipeptide boronic acid inhibitor bortezomib, the tetrapeptidic α′,β′epoxyketone carfilzomib and the tripeptide oprozomib are shown. All three compounds target the β5c and β5i active sites of the proteasome. Bortezomib and carfilzomib are FDA approved for the treatment of multiple myeloma; oprozomib is in phase I clinical trials. The unprimed substrate specificity (S) pockets of the CP that are targeted by these compounds are indicated

the long-term effects of bortezomib treatment revealed several adaptive mutations in the β5 substrate binding channel that convey drug resistance. For instance, the mutations M45V, M45I, A49T, A49V, A50V, C52F and C63F are suggested to impair both the catalytic activity of subunit β5 and its affinity for bortezomib, hereby leading to the observed reduction in therapeutic efficacy [93–96]. These findings might also provide an explanation for the up-regulation of subunit β5 as a compensation for its decreased activity [93]. However, evidence for the clinical relevance of these *in vitro* identified mutations and their biological impact still has to be adduced. Despite its overwhelming success in the first decade of its application, bortezomib's disadvantages encouraged the development of a second generation of CP inhibitors.

Just recently, in July 2012, one of these novel compounds, carfilzomib (Kyprolis®; Onyx Pharmaceuticals, Inc.; PR-171) has been approved by the FDA as a second-line drug for the treatment of patients that relapsed from bortezomib (Fig. 1.6) [97]. Carfilzomib belongs to the class of α′,β′epoxyketones derived from the microbial natural product epoxomicin. To date, due to their bivalent irreversible reaction mode, α′,β′epoxyketones are the most specific inhibitors known for the CP [81, 82] (Sect 4.3.4). Similarly to bortezomib, carfilzomib potently blocks the β5c and β5i active sites of cCP and iCP with IC_{50} values of 6 nM and 33 nM, respectively, but unlike bortezomib, carfilzomib does not affect β1c [98]. Although carfilzomib also unintentionally induces neutropenia and thrombocytopenia, it does not cause peripheral neurotoxicity, as observed for the boronic acid bortezomib [89, 99].

Oprozomib (Onyx Pharmaceuticals, Inc.; ONX 0912; PR-047), another epoxomicin derivative, is an orally available inhibitor of both the β5c and the β5i active sites (IC_{50} (β5c) 36 nM; IC_{50} (β5i) 82 nM) [100]. Owing to its high cytotoxicity and its superiority to bortezomib and carfilzomib with respect to its mode of administration, oprozomib is currently being explored in Phase I clinical trials as a monotherapy for solid tumours and haematological malignancies.

1.6 Subunit-Specific Proteasome Inhibitors and Their Therapeutic Potential

Subunit-specific inhibitory compounds are valuable tools for examining the impact of individual proteasome subunits on cell division and biological signalling pathways. However, the development of compounds with subunit specificities is often hindered by the strong inhibitory potency of most reactive functional head groups (see Sect. 1.4). Hence, only ligands that undergo optimal enthalpic interactions with the surrounding protein residues exert pronounced subunit selectivity and for their design structural data are strongly demanded.

A long time drug design efforts solely concentrated on the ChTL activities of the proteasome, but current studies also take into account the CL and TL active sites. Recently, a series of α′,β′epoxyketone inhibitors that target the TL activities

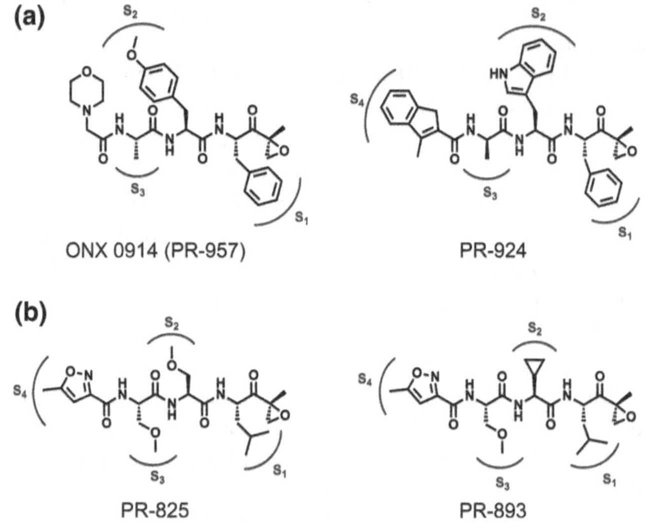

Fig. 1.7 Chemical structures of β5i- and β5c-selective compounds. (**a**) ONX 0914 and PR-924 specifically target subunit β5i of the iCP. (**b**) PR-825 and PR-893 are selective inhibitors of the β5c active site of the cCP. The unprimed substrate specificity pockets of the proteasome that accommodate the side chains of the peptidic inhibitors are indicated

of the cCP and iCP was published [101]. These compounds were demonstrated to render malignant cells more susceptible to bortezomib and carfilzomib than β1-selective inhibitors [101]. Nonetheless, compounds with selectivity for either β2c or β2i are still not available. For subunit β1i two selective drugs have been reported so far: UK-101, a dihydroeponemycin based α′,β′epoxyketone and IPSI-001 (calpeptin), a peptide aldehyde inhibitor. The cytotoxicity of UK-101 towards β1i overexpressing prostate cancer cells [60] underlines the importance of the iCP for the onset and progression of at least certain malignancies. Likewise, anti-tumour activity has been described for IPSI-001 [73], however, its apoptotic effects might result from co-inhibition of β5 active sites. For the β5 subunits several highly potent and selective ligands are known. The most potent compound is the β5i-targeting α′,β′epoxyketone ONX 0914 (PR-957; Onyx Pharmaceuticals, Inc; Fig. 1.7) [47]. Remarkably, ONX 0914 showed therapeutic effects in mouse models of rheumatoid arthritis [47], experimental colitis [71], lupus erythematosus [70] and Hashimoto's thyroiditis [72]. These medicinal benefits are based on the reduction of the levels of proinflammatory cytokines such as IL-6, IL-23 and TNF [47] and the modulation of CTL responses. More specifically, ONX 0914 harms the differentiation and proliferation of T_H1 and T_H17 cells but not of regulatory T lymphocytes [102]. In addition, administration of ONX 0914 depresses the expression of MHC I receptors to 50 % and abolishes the production of β5i-restricted antigens. Compared to etanercept, a scavenger of the proinflammatory cytokine TNF-α currently used as an immunosuppressant, ONX 0914 exerts higher potency [47]. Based on its promising pharmacological properties and effects in model systems, ONX 0914 is further investigated in preclinical studies for autoimmune diseases. The medicinal potential of the second β5i-selective compound PR-924 (Onyx Pharmaceuticals, Inc.), a structural analogue of ONX0914, has not yet been evaluated in immunological disorders, but conflicting data have been described with respect to its anti-myeloma activity [74, 75]. Notably, the two epoxomicin derivatives PR-825 and PR-893 (Onyx Pharmaceuticals, Inc.), that selectively target subunit β5c, were attributed neither anti-inflammatory nor anti-cancer activity [47, 75]. Hence, cCP and tCP selective drugs remain to be examined for any medicinal use.

References

1. D. Finley, Recognition and processing of ubiquitin-protein conjugates by the proteasome. Ann. Rev. Biochem. **78**, 477–513 (2009)
2. A.L. Goldberg, K.L. Rock, Proteolysis, proteasomes and antigen presentation. Nature **357**, 375–379 (1992)
3. J.D. Etlinger, A.L. Goldberg, A soluble ATP-dependent proteolytic system responsible for the degradation of abnormal proteins in reticulocytes. Proc. Natl. Acad. Sci. U.S.A **74**, 54–58 (1977)
4. A. Hershko, A. Ciechanover, The ubiquitin system. Ann. Rev. Biochem. **67**, 425–479 (1998)
5. I.A. York, S.C. Chang, T. Saric, J.A. Keys, J.M. Favreau, A.L. Goldberg, K.L. Rock, The ER aminopeptidase ERAP1 enhances or limits antigen presentation by trimming epitopes to 8-9 residues. Nat. Immunol. **3**, 1177–1184 (2002)
6. K.L. Rock, I.A. York, A.L. Goldberg, Post-proteasomal antigen processing for major histocompatibility complex class I presentation. Nat. Immunol. **5**, 670–677 (2004)
7. A. Townsend, J. Trowsdale, The transporters associated with antigen presentation. Semin. Cell Biol. **4**, 53–61 (1993)
8. V.H. Engelhard, Structure of peptides associated with MHC class I molecules. Curr. Opin. Immunol. **6**, 13–23 (1994)
9. K. Falk, O. Rotzschke, S. Stevanovic, G. Jung, H.G. Rammensee, Allele-specific motifs revealed by sequencing of self-peptides eluted from MHC molecules. Nature **351**, 290–296 (1991)
10. J.M. Vyas, A.G. Van der Veen, H.L. Ploegh, The known unknowns of antigen processing and presentation. Nat. Rev. Immunol. **8**, 607–618 (2008)
11. J. Neefjes, M.L. Jongsma, P. Paul, O. Bakke, Towards a systems understanding of MHC class I and MHC class II antigen presentation. Nat. Rev. Immunol. **11**, 823–836 (2011)
12. C.M. Pickart, Mechanisms underlying ubiquitination. Ann. Rev. Biochem. **70**, 503–533 (2001)
13. J.M. Berg, J.L. Tymoczko, L. Stryer, *Biochemistry* (W H Freeman, New York, 2002)
14. D. Voges, P. Zwickl, W. Baumeister, The 26S proteasome: a molecular machine designed for controlled proteolysis. Ann. Rev. Biochem. **68**, 1015–1068 (1999)
15. M. Bochtler, L. Ditzel, M. Groll, R. Huber, Crystal structure of heat shock locus V (HslV) from Escherichia coli. Proc. Natl. Acad. Sci. U.S.A **94**, 6070–6074 (1997)
16. J. Löwe, D. Stock, B. Jap, P. Zwickl, W. Baumeister, R. Huber, Crystal structure of the 20S proteasome from the archaeon *T. acidophilum* at 3.4 Å resolution. Science **268**, 533–539 (1995)
17. M. Groll, L. Ditzel, J. Löwe, D. Stock, M. Bochtler, H.D. Bartunik, R. Huber, Structure of 20S proteasome from yeast at 2.4 Å resolution. Nature **386**, 463–471 (1997)

18. P. Zwickl, A. Grziwa, G. Puhler, B. Dahlmann, F. Lottspeich, W. Baumeister, Primary structure of the Thermoplasma proteasome and its implications for the structure, function, and evolution of the multicatalytic proteinase. Biochemistry **31**, 964–972 (1992)
19. M. Unno, T. Mizushima, Y. Morimoto, Y. Tomisugi, K. Tanaka, N. Yasuoka, T. Tsukihara (2002). The structure of the mammalian 20S proteasome at 2.75 Å resolution, Structure, 10, 609-618
20. M. Groettrup, R. Kraft, S. Kostka, S. Standera, R. Stohwasser, P.M. Kloetzel, A third interferon-gamma-induced subunit exchange in the 20S proteasome. Eur. J. Immunol. **26**, 863–869 (1996)
21. S. Murata, K. Sasaki, T. Kishimoto, S. Niwa, H. Hayashi, Y. Takahama, K. Tanaka, Regulation of CD8+ T cell development by thymus-specific proteasomes. Science **316**, 1349–1353 (2007)
22. E.M. Huber, M. Groll, Inhibitors for the immuno- and constitutive proteasome: current and future trends in drug development. Angew. Chem. Int. Ed. Engl. **51**, 8708–8720 (2012)
23. K. Tanaka, T. Yoshimura, T. Tamura, T. Fujiwara, A. Kumatori, A. Ichihara, Possible mechanism of nuclear translocation of proteasomes. FEBS Lett. **271**, 41–46 (1990)
24. M. Groll, M. Bajorek, A. Köhler, L. Moroder, D.M. Rubin, R. Huber, M.H. Glickman, D. Finley, A gated channel into the proteasome core particle. Nat. Struct. Biol. **7**, 1062–1067 (2000)
25. F.G. Whitby, E.I. Masters, L. Kramer, J.R. Knowlton, Y. Yao, C.C. Wang, C.P. Hill, Structural basis for the activation of 20S proteasomes by 11S regulators. Nature **408**, 115–120 (2000)
26. B.M. Stadtmueller, C.P. Hill, Proteasome activators. Mol. Cell **41**, 8–19 (2011)
27. P.C. Ramos, J. Hockendorff, E.S. Johnson, A. Varshavsky, R.J. Dohmen, Ump1p is required for proper maturation of the 20S proteasome and becomes its substrate upon completion of the assembly. Cell **92**, 489–499 (1998)
28. M. Groll, W. Heinemeyer, S. Jäger, T. Ullrich, M. Bochtler, D.H. Wolf, R. Huber, The catalytic sites of 20S proteasomes and their role in subunit maturation: a mutational and crystallographic study. Proc. Natl. Acad. Sci. U.S.A **96**, 10976–10983 (1999)
29. L. Ditzel, R. Huber, K. Mann, W. Heinemeyer, D.H. Wolf, M. Groll, Conformational constraints for protein self-cleavage in the proteasome. J. Mol. Biol. **279**, 1187–1191 (1998)
30. M. Groettrup, C.J. Kirk, M. Basler, Proteasomes in immune cells: more than peptide producers? Nat. Rev. Immunol. **10**, 73–78 (2010)
31. M. Aki, N. Shimbara, M. Takashina, K. Akiyama, S. Kagawa, T. Tamura, N. Tanahashi, T. Yoshimura, K. Tanaka, A. Ichihara, Interferon-gamma induces different subunit organizations and functional diversity of proteasomes. J. Biochem. **115**, 257–269 (1994)
32. T.A. Griffin, D. Nandi, M. Cruz, H.J. Fehling, L.V. Kaer, J.J. Monaco, R.A. Colbert, Immunoproteasome assembly: cooperative incorporation of interferon gamma (IFN-gamma)-inducible subunits. J. Exp. Med. **187**, 97–104 (1998)
33. C. Realini, W. Dubiel, G. Pratt, K. Ferrell, M. Rechsteiner, Molecular cloning and expression of a gamma-interferon-inducible activator of the multicatalytic protease. J. Biol. Chem. **269**, 20727–20732 (1994)
34. M. Rechsteiner, C.P. Hill, Mobilizing the proteolytic machine: cell biological roles of proteasome activators and inhibitors. Trends Cell Biol. **15**, 27–33 (2005)
35. M. Rechsteiner, C. Realini, V. Ustrell, The proteasome activator 11 S REG (PA28) and class I antigen presentation. Biochem. J. **345**(Pt 1), 1–15 (2000)
36. B. Guillaume, J. Chapiro, V. Stroobant, D. Colau, B. Van Holle, G. Parvizi, M.P. Bousquet-Dubouch, I. Theate, N. Parmentier, B.J. Van den Eynde, Two abundant proteasome subtypes that uniquely process some antigens presented by HLA class I molecules. Proc. Natl. Acad. Sci. U.S.A **107**, 18599–18604 (2010)
37. N. Klare, M. Seeger, K. Janek, P.R. Jungblut, B. Dahlmann, Intermediate-type 20 S proteasomes in HeLa cells: "asymmetric" subunit composition, diversity and adaptation. J. Mol. Biol. **373**, 1–10 (2007)
38. M. Groll, M. Bochtler, H. Brandstetter, T. Clausen, R. Huber, Molecular machines for protein degradation. ChemBioChem **6**, 222–256 (2005)

39. E. Huber, M. Basler, R. Schwab, W. Heinemeyer, C.J. Kirk, M. Groettrup, M. Groll, Immuno- and constitutive proteasome crystal structures reveal differences in substrate and inhibitor specificity. Cell **148**, 727–738 (2012)
40. C. Cardozo, R.A. Kohanski, Altered properties of the branched chain amino acid-preferring activity contribute to increased cleavages after branched chain residues by the "immunoproteasome". J. Biol. Chem. **273**, 16764–16770 (1998)
41. M. Orlowski, C. Cardozo, C. Michaud, Evidence for the presence of five distinct proteolytic components in the pituitary multicatalytic proteinase complex. Properties of two components cleaving bonds on the carboxyl side of branched chain and small neutral amino acids. Biochemistry **32**, 1563–1572 (1993)
42. M. Gaczynska, K.L. Rock, A.L. Goldberg, Gamma-interferon and expression of MHC genes regulate peptide hydrolysis by proteasomes. Nature **365**, 264–267 (1993)
43. J. Driscoll, M.G. Brown, D. Finley, J.J. Monaco, MHC-linked LMP gene products specifically alter peptidase activities of the proteasome. Nature **365**, 262–264 (1993)
44. B.J. Van den Eynde, S. Morel, Differential processing of class-I-restricted epitopes by the standard proteasome and the immunoproteasome. Curr. Opin. Immunol. **13**, 147–153 (2001)
45. S. Murata, Y. Takahama, K. Tanaka, Thymoproteasome: probable role in generating positively selecting peptides. Curr. Opin. Immunol. **20**, 192–196 (2008)
46. E.Z. Kincaid, J.W. Che, I. York, H. Escobar, E. Reyes-Vargas, J.C. Delgado, R.M. Welsh, M.L. Karow, A.J. Murphy, D.M. Valenzuela, G.D. Yancopoulos, K.L. Rock, Mice completely lacking immunoproteasomes show major changes in antigen presentation. Nat. Immunol. **13**, 129–135 (2011)
47. T. Muchamuel, M. Basler, M. A. Aujay, E. Suzuki, K.W. Kalim, C. Lauer, C. Sylvain, E.R. Ring, J. Shields, J. Jiang, P. Shwonek, F. Parlati, S.D. Demo, M.K. Bennett, C.J. Kirk, M. Groettrup, A selective inhibitor of the immunoproteasome subunit LMP7 blocks cytokine production and attenuates progression of experimental arthritis. Nat. Med. **15**, 781–787 (2009)
48. U. Seifert, L.P. Bialy, F. Ebstein, D. Bech-Otschir, A. Voigt, F. Schroter, T. Prozorovski, N. Lange, J. Steffen, M. Rieger, U. Kuckelkorn, O. Aktas, P.M. Kloetzel, E. Kruger, Immunoproteasomes preserve protein homeostasis upon interferon-induced oxidative stress. Cell **142**, 613–624 (2010)
49. N.R. Gascoigne, E. Palmer, Signaling in thymic selection. Curr. Opin. Immunol. **23**, 207–212 (2011)
50. M.J. Bevan, Immunology. The cutting edge of T cell selection. Science **316**, 1291–1292 (2007)
51. P.M. Voorhees, E.C. Dees, B. O'Neil, R.Z. Orlowski, The proteasome as a target for cancer therapy. Clin. Cancer Res. **9**, 6316–6325 (2003)
52. L.R. Dick, P.E. Fleming, Building on bortezomib: second-generation proteasome inhibitors as anti-cancer therapy. Drug Discov. Today **15**, 243–249 (2010)
53. D.J. McConkey, K. Zhu, Mechanisms of proteasome inhibitor action and resistance in cancer. Drug Resist. Updat. **11**, 164–179 (2008)
54. S.T. Nawrocki, J.S. Carew, K. Dunner Jr., L.H. Boise, P.J. Chiao, P. Huang, J.L. Abbruzzese, D.J. McConkey, Bortezomib inhibits PKR-like endoplasmic reticulum (ER) kinase and induces apoptosis via ER stress in human pancreatic cancer cells. Cancer Res. **65**, 11510–11519 (2005)
55. E.A. Obeng, L.M. Carlson, D.M. Gutman, W.J. Harrington Jr., K.P. Lee, L.H. Boise, Proteasome inhibitors induce a terminal unfolded protein response in multiple myeloma cells. Blood **107**, 4907–4916 (2006)
56. G. Bianchi, L. Oliva, P. Cascio, N. Pengo, F. Fontana, F. Cerruti, A. Orsi, E. Pasqualetto, A. Mezghrani, V. Calbi, G. Palladini, N. Giuliani, K.C. Anderson, R. Sitia, S. Cenci, The proteasome load versus capacity balance determines apoptotic sensitivity of multiple myeloma cells to proteasome inhibition. Blood **113**, 3040–3049 (2009)
57. T. Mujtaba, Q.P. Dou, Advances in the understanding of mechanisms and therapeutic use of bortezomib. Discov. Med. **12**, 471–480 (2011)

58. I. Nickeleit, S. Zender, F. Sasse, R. Geffers, G. Brandes, I. Sorensen, H. Steinmetz, S. Kubicka, T. Carlomagno, D. Menche, I. Gutgemann, J. Buer, A. Gossler, M.P. Manns, M. Kalesse, R. Frank, N.P. Malek, Argyrin a reveals a critical role for the tumor suppressor protein p27(kip1) in mediating antitumor activities in response to proteasome inhibition. Cancer Cell **14**, 23–35 (2008)
59. W. Lee, K.B. Kim, The immunoproteasome: an emerging therapeutic target. Curr. Top. Med. Chem. **11**, 2923–2930 (2011)
60. Y.K. Ho, P. Bargagna-Mohan, M. Wehenkel, R. Mohan, K.B. Kim, LMP2-specific inhibitors: chemical genetic tools for proteasome biology. Chem. Biol. **14**, 419–430 (2007)
61. M. Mishto, E. Bellavista, A. Santoro, A. Stolzing, C. Ligorio, B. Nacmias, L. Spazzafumo, M. Chiappelli, F. Licastro, S. Sorbi, A. Pession, T. Ohm, T. Grune, C. Franceschi, Immunoproteasome and LMP2 polymorphism in aged and Alzheimer's disease brains. Neurobiol. Aging **27**, 54–66 (2006)
62. M. Diaz-Hernandez, F. Hernandez, E. Martin-Aparicio, P. Gomez-Ramos, M.A. Moran, J.G. Castano, I. Ferrer, J. Avila, J.J. Lucas, Neuronal induction of the immunoproteasome in Huntington's disease. J. Neurosci. **23**, 11653–11661 (2003)
63. K. Puttaparthi, J.L. Elliott, Non-neuronal induction of immunoproteasome subunits in an ALS model: possible mediation by cytokines. Exp. Neurol. **196**, 441–451 (2005)
64. L.R. Fitzpatrick, J.S. Small, L.S. Poritz, K.J. McKenna, W.A. Koltun, Enhanced intestinal expression of the proteasome subunit low molecular mass polypeptide 2 in patients with inflammatory bowel disease. Dis. Colon Rectum **50**, 337–348; discussion 348–350 (2007)
65. T. Egerer, L. Martinez-Gamboa, A. Dankof, B. Stuhlmuller, T. Dorner, V. Krenn, K. Egerer, P.E. Rudolph, G.R. Burmester, E. Feist, Tissue-specific up-regulation of the proteasome subunit beta5i (LMP7) in Sjogren's syndrome. Arthritis Rheum. **54**, 1501–1508 (2006)
66. I. Ferrer, B. Martin, J.G. Castano, J.J. Lucas, D. Moreno, M. Olive, Proteasomal expression, induction of immunoproteasome subunits, and local MHC class I presentation in myofibrillar myopathy and inclusion body myositis. J. Neuropathol. Exp. Neurol. **63**, 484–498 (2004)
67. Z. Yang, D. Gagarin, G. St Laurent 3rd, N. Hammell, I. Toma, C.A. Hu, A. Iwasa, T.A. McCaffrey, Cardiovascular inflammation and lesion cell apoptosis: a novel connection via the interferon-inducible immunoproteasome. Arterioscler. Thromb. Vasc. Biol. **29**, 1213–1219 (2009)
68. A. Visekruna, N. Slavova, S. Dullat, J. Grone, A.J. Kroesen, J.P. Ritz, H.J. Buhr, U. Steinhoff, Expression of catalytic proteasome subunits in the gut of patients with Crohn's disease. Int. J. Colorectal Dis. **24**, 1133–1139 (2009)
69. L.R. Fitzpatrick, V. Khare, J.S. Small, W.A. Koltun, Dextran sulfate sodium-induced colitis is associated with enhanced low molecular mass polypeptide 2 (LMP2) expression and is attenuated in LMP2 knockout mice. Dig. Dis. Sci. **51**, 1269–1276 (2006)
70. H.T. Ichikawa, T. Conley, T. Muchamuel, J. Jiang, S. Lee, T. Owen, J. Barnard, S. Nevarez, B.I. Goldman, C.J. Kirk, R.J. Looney, J.H. Anolik, Novel proteasome inhibitors have a beneficial effect in murine lupus via the dual inhibition of type i interferon and autoantibody secreting cells. Arthritis Rheum. **64**, 493–503 (2011)
71. M. Basler, M. Dajee, C. Moll, M. Groettrup, C.J. Kirk, Prevention of experimental colitis by a selective inhibitor of the immunoproteasome. J. Immunol. **185**, 634–641 (2010)
72. Y. Nagayama, M. Nakahara, M. Shimamura, I. Horie, K. Arima, N. Abiru, Prophylactic and therapeutic efficacies of a selective inhibitor of the immunoproteasome for Hashimoto's thyroiditis, but not for Graves' hyperthyroidism, in mice. Clin. Exp. Immunol. **168**, 268–273 (2012)
73. D.J. Kuhn, S.A. Hunsucker, Q. Chen, P.M. Voorhees, M. Orlowski, R.Z. Orlowski, Targeted inhibition of the immunoproteasome is a potent strategy against models of multiple myeloma that overcomes resistance to conventional drugs and nonspecific proteasome inhibitors. Blood **113**, 4667–4676 (2009)
74. A.V. Singh, M. Bandi, M.A. Aujay, C.J. Kirk, D.E. Hark, N. Raje, D. Chauhan, K.C. Anderson, PR-924, a selective inhibitor of the immunoproteasome subunit LMP-7, blocks multiple myeloma cell growth both in vitro and in vivo. Br. J. Haematol. **152**, 155–163 (2011)

75. F. Parlati, S.J. Lee, M. Aujay, E. Suzuki, K. Levitsky, J.B. Lorens, D.R. Micklem, P. Ruurs, C. Sylvain, Y. Lu, K.D. Shenk, M.K. Bennett, Carfilzomib can induce tumor cell death through selective inhibition of the chymotrypsin-like activity of the proteasome. Blood **114**, 3439–3447 (2009)
76. A. Vinitsky, C. Michaud, J.C. Powers, M. Orlowski, Inhibition of the chymotrypsin-like activity of the pituitary multicatalytic proteinase complex. Biochemistry **31**, 9421–9428 (1992)
77. M. Bogyo, J.S. McMaster, M. Gaczynska, D. Tortorella, A.L. Goldberg, H. Ploegh, Covalent modification of the active site threonine of proteasomal beta subunits and the Escherichia coli homolog HslV by a new class of inhibitors. Proc. Natl. Acad. Sci. U.S.A **94**, 6629–6634 (1997)
78. M. Groll, B. Schellenberg, A.S. Bachmann, C.R. Archer, R. Huber, T.K. Powell, S. Lindow, M. Kaiser, R. Dudler, A plant pathogen virulence factor inhibits the eukaryotic proteasome by a novel mechanism. Nature **452**, 755–758 (2008)
79. M. Groll, R. Huber, B.C. Potts, Crystal structures of Salinosporamide A (NPI-0052) and B (NPI-0047) in complex with the 20S proteasome reveal important consequences of beta-lactone ring opening and a mechanism for irreversible binding. J. Am. Chem. Soc. **128**, 5136–5141 (2006)
80. M. Groll, B.C. Potts, Proteasome structure, function, and lessons learned from beta-lactone inhibitors. Curr. Top. Med. Chem. **11**, 2850–2878 (2011)
81. M. Groll, K.B. Kim, N. Kairies, R. Huber, C.M. Crews, Crystal structure of epoxomicin: 20S proteasome reveals a molecular basis for selectivity of α', β'-epoxyketone proteasome inhibitors. J. Am. Chem. Soc. **122**, 1237–1238 (2000)
82. K.B. Kim, J. Myung, N. Sin, C.M. Crews, Proteasome inhibition by the natural products epoxomicin and dihydroeponemycin: insights into specificity and potency. Bioorg. Med. Chem. Lett. **9**, 3335–3340 (1999)
83. M.A. Gräwert, N. Gallastegui, M. Stein, B. Schmidt, P.M. Kloetzel, R. Huber, M. Groll, Elucidation of the alpha-keto-aldehyde binding mechanism: a lead structure motif for proteasome inhibition. Angew. Chem. Int. Ed. Engl. **50**, 542–544 (2011)
84. L. Borissenko, M. Groll, 20S proteasome and its inhibitors: crystallographic knowledge for drug development. Chem. Rev. **107**, 687–717 (2007)
85. R.C. Kane, P.F. Bross, A.T. Farrell, R. Pazdur, Velcade: U.S. FDA approval for the treatment of multiple myeloma progressing on prior therapy. Oncologist **8**, 508–513 (2003)
86. I. Tzvetanov, M. Spaggiari, J. Joseph, H. Jeon, J. Thielke, J. Oberholzer, E. Benedetti, The use of bortezomib as a rescue treatment for acute antibody-mediated rejection: report of three cases and review of literature. Transplant. Proc. **44**, 2971–2975 (2012)
87. M. Escobar, M. Velez, A. Belalcazar, E.S. Santos, L.E. Raez, The role of proteasome inhibition in nonsmall cell lung cancer. J. Biomed. Biotechnol. **2011**, 806506 (2011)
88. M. Groll, C.R. Berkers, H.L. Ploegh, H. Ovaa, Crystal structure of the boronic acid-based proteasome inhibitor bortezomib in complex with the yeast 20S proteasome. Structure **14**, 451–456 (2006)
89. S. Arastu-Kapur, J.L. Anderl, M. Kraus, F. Parlati, K.D. Shenk, S.J. Lee, T. Muchamuel, M.K. Bennett, C. Driessen, A.J. Ball, C.J. Kirk, Nonproteasomal targets of the proteasome inhibitors bortezomib and carfilzomib: a link to clinical adverse events. Clin. Cancer Res. **17**, 2734–2743 (2011)
90. P. Moreau, P.G. Richardson, M. Cavo, R.Z. Orlowski, J.F. San Miguel, A. Palumbo, J.L. Harousseau, Proteasome inhibitors in multiple myeloma: 10 years later. Blood **120**, 947–959 (2012)
91. S. Kumar, S.V. Rajkumar, Many facets of bortezomib resistance/susceptibility. Blood **112**, 2177–2178 (2008)
92. D. Chauhan, G. Li, R. Shringarpure, K. Podar, Y. Ohtake, T. Hideshima, K.C. Anderson, Blockade of Hsp27 overcomes Bortezomib/proteasome inhibitor PS-341 resistance in lymphoma cells. Cancer Res. **63**, 6174–6177 (2003)
93. R. Oerlemans, N.E. Franke, Y.G. Assaraf, J. Cloos, I. van Zantwijk, C.R. Berkers, G.L. Scheffer, K. Debipersad, K. Vojtekova, C. Lemos, J.W. van der Heijden, B. Ylstra, G.J.

Peters, G.L. Kaspers, B.A. Dijkmans, R.J. Scheper, G. Jansen, Molecular basis of bortezomib resistance: proteasome subunit beta5 (PSMB5) gene mutation and overexpression of PSMB5 protein. Blood **112**, 2489–2499 (2008)

94. S. Lu, J. Yang, Z. Chen, S. Gong, H. Zhou, X. Xu, J. Wang, Different mutants of PSMB5 confer varying bortezomib resistance in T lymphoblastic lymphoma/leukemia cells derived from the Jurkat cell line. Exp. Hematol. **37**, 831–837 (2009)
95. N.E. Franke, D. Niewerth, Y.G. Assaraf, J. van Meerloo, K. Vojtekova, C.H. van Zantwijk, S. Zweegman, E.T. Chan, C.J. Kirk, D.P. Geerke, A.D. Schimmer, G.J. Kaspers, G. Jansen, J. Cloos, Impaired bortezomib binding to mutant beta5 subunit of the proteasome is the underlying basis for bortezomib resistance in leukemia cells. Leukemia **26**, 757–768 (2011)
96. E. Suzuki, S. Demo, E. Deu, J. Keats, S. Arastu-Kapur, P.L. Bergsagel, M.K. Bennett, C.J. Kirk, Molecular mechanisms of bortezomib resistant adenocarcinoma cells. PLoS One **6**, e27996 (2011)
97. K. Fostier, A. De Becker, R. Schots, Carfilzomib: a novel treatment in relapsed and refractory multiple myeloma. OncoTargets Ther. **5**, 237–244 (2012)
98. S.D. Demo, C.J. Kirk, M.A. Aujay, T.J. Buchholz, M. Dajee, M.N. Ho, J. Jiang, G.J. Laidig, E.R. Lewis, F. Parlati, K.D. Shenk, M.S. Smyth, C.M. Sun, M.K. Vallone, T.M. Woo, C.J. Molineaux, M.K. Bennett, Antitumor activity of PR-171, a novel irreversible inhibitor of the proteasome. Cancer Res. **67**, 6383–6391 (2007)
99. A.M. Ruschak, M. Slassi, L.E. Kay, A.D. Schimmer, Novel proteasome inhibitors to overcome bortezomib resistance. J. Natl Cancer Inst. **103**, 1007–1017 (2011)
100. H.J. Zhou, M.A. Aujay, M.K. Bennett, M. Dajee, S.D. Demo, Y. Fang, M.N. Ho, J. Jiang, C.J. Kirk, G.J. Laidig, E.R. Lewis, Y. Lu, T. Muchamuel, F. Parlati, E. Ring, K.D. Shenk, J. Shields, P.J. Shwonek, T. Stanton, C.M. Sun, C. Sylvain, T.M. Woo, J. Yang, Design and synthesis of an orally bioavailable and selective peptide epoxyketone proteasome inhibitor (PR-047). J. Med. Chem. **52**, 3028–3038 (2009)
101. A.C. Mirabella, A.A. Pletnev, S.L. Downey, B.I. Florea, T.B. Shabaneh, M. Britton, M. Verdoes, D.V. Filippov, H.S. Overkleeft, A.F. Kisselev, Specific cell-permeable inhibitor of proteasome trypsin-like sites selectively sensitizes myeloma cells to bortezomib and carfilzomib. Chem. Biol. **18**, 608–618 (2011)
102. K.W. Kalim, M. Basler, C.J. Kirk, M. Groettrup, Immunoproteasome Subunit LMP7 Deficiency and Inhibition Suppresses Th1 and Th17 but Enhances Regulatory T Cell Differentiation. J. Immunol. **189**, 4182–4193 (2012)

Chapter 2
Objective

Structural information on the yCP alone and in complex with ligands strongly supported the development of the currently known potent CP inhibitors bortezomib and carfilzomib. Both compounds do not discriminate between the three mammalian CP types and, thereby, exert anti-cancer activity. However, recent studies demonstrate that iCP-specific inhibitors might qualify as therapeutics in autoimmune diseases. So far, only a few iCP-specific inhibitors were identified, mostly because structural data on the iCP were lacking.

The aim of this thesis was to elucidate the atomic structures of the murine iCP and cCP by X-ray crystallography. The crystal structures of both CP types from one organism enable the direct comparison of the structural features of the iCP and the cCP from *Mus musculus*, the cCP from *B. taurus* and the CP of *S. cerevisiae*. The structural data were expected to explain observed differences in the substrate specificities of the iCP and cCP. Moreover the structural characterization aimed at providing insights into the generation of the distinct MHC I peptide patterns by the cCP and iCP and into the pivotal role of the iCP in the adaptive immune system of vertebrates. Besides the examination of the cleavage preferences, it was intended to elucidate the molecular basis for the (non-) selectivity of known CP inhibitors. In particular, cCP and iCP structures in complex with the iCP-specific epoxyketone inhibitor ONX 0914 were supposed to unravel the reason for its β5i selectivity.

Amino acid substitutions in the substrate binding channel are known to affect cleavage and inhibitor specificities. To evaluate the impact of amino acid differences between β5 subunits on the proteolytic activity mutagenesis experiments with the model organism *S. cerevisiae* were envisioned. These aimed at imitating the active site surroundings of subunit β5c, β5i and β5t. Structural data on these mutant yCPs and analysis of their affinity towards ONX 0914 and bortezomib were expected to reveal the effect of the amino acid substitutions on substrate and inhibitor specificities.

In summary, a multidisciplinary approach combining X-ray crystallography, yeast mutagenesis and inhibition assays was intended to provide detailed information on the architecture of all active sites of murine CP types for the development of potential lead structures for diverse medicinal applications.

E. M. Huber, *Structural and Functional Characterization of the Immunoproteasome*, Springer Theses, DOI: 10.1007/978-3-319-01556-9_2,

Chapter 3
Materials and Methods

3.1 Materials

3.1.1 Chemicals

All chemicals were obtained from the following companies:

AppliChem (Darmstadt, DE), Biomol (Hamburg, DE), Fluka (Neu-Ulm, DE), Merck (Darmstadt, DE), Sigma-Aldrich (Steinheim, DE), Serva (Heidelberg, DE), Roth (Karlsruhe, DE) and VWR (Darmstadt, DE).

3.1.2 Antibiotics

Ampicillin was used in a final concentration of 180 mg/l for the selection of transformed *Escherichia coli* (*E. coli*).

3.1.3 Media

LB_0 medium	Peptone	1 % (w/v)
	Yeast extract	0.5 % (w/v)
	NaCl	0.5 % (w/v)
	Agar	2 % (w/v)
SOC medium	Peptone	2 % (w/v)
	Yeast extract	0.5 % (w/v)
	Glucose	20 mM
	$MgSO_4$	10 mM
	NaCl	10 mM
	$MgCl_2$	10 mM
	KCl	2.5 mM

(continued)

E. M. Huber, *Structural and Functional Characterization of the Immunoproteasome*, Springer Theses, DOI: 10.1007/978-3-319-01556-9_3,

(continued)

YPD medium	Yeast extract	1 % (w/v)
	Peptone	2 % (w/v)
	Glucose	2 % (w/v)
	Agar	2 % (w/v)
Synthetic complete medium (CM)	Dropout powder	0.13 % (w/v)
	Yeast nitrogen base w/o aa	0.67 % (w/v)
	Glucose	2 % (w/v)
	pH 5.6	
CM medium ura$^-$ leu$^-$ his$^-$	2x CM medium	300 ml
	4 % (w/v) Agar	300 ml
	40 % (w/v) Glucose	30 ml
	Adenine (30 mM)	6 ml
	Tryptophane (40 mM)	6 ml
	Lysine (100 mM)	6 ml
CM medium leu$^-$ his$^-$ 5-FOA	2x CM medium	300 ml
	4 % (w/v) Agar	300 ml
	40 % (w/v) Glucose	30 ml
	Adenine (30 mM)	6 ml
	Tryptophane (40 mM)	6 ml
	Lysine (100 mM)	6 ml
	Uracil (20 mM)	15 ml
	5-fluoroorotic acid (5-FOA)	0.6 g

3.1.4 Enzymes

Phusion DNA polymerase (2 U/µl)	Finnzymes (Vantaa, FI)
Pfu Turbo DNA polymerase®	Agilent (Santa Clara, US)
Restriction endonuclease *Bam*HI (20 U/µl)	New England biolabs (Ipswich, US)
Restriction endonuclease *Hin*dIII (20 U/µl)	New England biolabs (Ipswich, US)
Restriction endonuclease *Dpn*I (20 U/µl)	New England biolabs (Ipswich, US)
T4 DNA ligase (1 U/µl)	Invitrogen (Carlsbad, US)
DNAse I	Sigma-Aldrich (St. Louis, US)

3.1.5 Primer

All oligonucleotides were either HPSF or HPLC purified and dissolved in ddH_2O to a final concentration of 100 pmol/µl. The working concentration of the primer was 10 pmol/µl. All primer were synthesized by Eurofins MWG Operon, Ebersberg (Table 3.1).

Table 3.1 Oligonucleotides used in this work

Primer	Sequence 5′ → 3′
5/6F	GAAACAGCTATGACCATGAT
5/6R	GACGGCCAGTGAATTGTAAT
20S_22C_for	AGATTCTCGTT/GCCACTTGTGGCAATTC
20S_22C_rev	CAATTGCCACAAGTGGA/CACGAGAATCT
27S_31M_rev	CTTCAC/TAGTTTGAGAAGC/AAACCCAATTGC
31M_for	CTTCTCAAACTG/ATGAAGAAAGTTATTG
31M_for2	CTTCTCAAACTATGAAGAAAGTTATTGAG
31S_for	TCTCAAACTTCTAAGAAAGTTATTG
31S_rev	CAATAACTTTCTTAGAAGTTTGAGA
45T_48T_for	TGGGTACAACTTCCGGTACTGCGGCAG
45T_48T_rev	CTGCCGCAGTACCGGAAGTTGTACCCA
57R_for	TTTTGGGAACGTTGGCTAGGT
57R_rev	ACCTAGCCAACGTTCCCAAAA
32N_for	CTCAAACTG/ATGAATAAAGTTATTG
32N_rev	CAATAACTTTATTCAT/CAGTTTGAG
53S_for	GGCAGATTGTTCATTTTGGG
53S_rev	CCCAAAATGAACAATCTGCC
53S_57R-for	GTTCATTTTGGGAACGTTGGCTAGGTTCT
53S_57R-rev	CAACGTTCCCAAAATGAACAATCTGCCGC
SCSTRNG-M45A_for	CATTTTTATTGGGTACCGCGTCTGGTTGTGCGGCAG
SCSTRNG-M45A_rev	CTGCCGCACAACCAGACGCGGTACCCAATAAAAATG
SCSTRNG-M45V_for	CATTTTTATTGGGTACCGTTTCTGGTTGTGCGGCAG
SCSTRNG-M45V_rev	CTGCCGCACAACCAGAAACGGTACCCAATAAAAATG
71G_for	CGAGCTGAGGGAAGGTGAACGTATATCGC
71G_rev	GCGATATACGTTCACCTTCCCTCAGCTCG
127T_for	GACATATTCTGCACTGGTTCAGGTCAA
127T_rev	TTGACCTGAACCAGTGCAGAATATGTC

Underlined nucleotides encode mutations

3.1.6 Strains

The *E. coli* strains XL1-Blue and DH5α were used for cloning of mutant *pre2* genes in the vector pRS315. The created plasmids were introduced in *S. cerevisiae* YWH20a. *S. cerevisiae* WCG4a and WCG4a *pre1-1* served as control strains in proteasome activity tests (see Sect. 3.4.1 and Table 3.2).

3.1.7 Plasmids

pRS315

The plasmid pRS315 is a pBluescript-based centromere vector composed of 6 kbp. The plasmid encodes the enzyme β lactamase, thereby conferring resistance

Table 3.2 *E. coli* and yeast strains used in this work

Organism	Strain	Genotype	Source
E. coli	XL1-Blue	*recA1, endA1, gyrA96, thi-1, hsdR17, supE44, relA1, lac,* [F′, *proAB, lacI^qZΔM15,* Tn*10* (Tet^r)]	Bullock [1]
E. coli	DH5α	*fhuA2 Δ(argF-lacZ)U169 phoA glnV44 Φ80 Δ(lacZ)M15 gyrA96 recA1 relA1 endA1 thi-1 hsdR17*	Hanahan [2]
S. cerevisiae	WCG4a	*MAT*a *leu2-3,112 ura3 his3-11,15 Can^S GAL2*	Heinemeyer [3]
S. cerevisiae	WCG4a *pre1-1*	*MAT*a *leu2-3,112 ura3 his3-11,15 rad5-535 Can^S GAL2 pre1-1 (S142F)*	Heinemeyer [4]
S. cerevisiae	YWH20a	*MAT*a *leu2-3,112 ura3 his3-11,15 Can^S GAL2 pre2Δ::HIS3* [pRS316-E2]	Heinemeyer [5]

to ampicillin and enabling selection in *E. coli*. The yeast selectable marker gene is *LEU2*. *PRE2* mutant genes of 1.5 kbp were inserted via the restriction sites *Bam*HI and *Hin*dIII into the multiple cloning site (see Table 3.3). Expression of the *pre2* mutants was controlled by the endogenic *PRE2* promoter.

pRS316-E2

The plasmid pRS316 is pBluescript-based centromere vector composed of 4.9 kbp encoding the ampicillin resistance gene and the *URA3* selection marker, suitable for replication in both *E. coli* and yeast. Expression of the wildtype (wt) *PRE2* gene, inserted via the restriction sites *Bam*HI and *Hin*dIII, was under the control of the endogenic *PRE2* promoter.

3.1.8 Fluorogenic Substrates

Cbz-Gly-Gly-Leu-pNA	Bachem (Bubendorf, Switzerland)
Suc-Leu-Leu-Val-Tyr-AMC	Bachem (Bubendorf, Switzerland)
Cbz-Leu-Leu-Glu-AMC	Bachem (Bubendorf, Switzerland)

All fluorogenic substrates were dissolved in DMSO and stored at −20 °C.

3.1.9 DNA and Protein Standards

peqGOLD DNA ladder mix (100–10,000 bp)	Peqlab (Erlangen, DE)
Roti®-Mark STANDARD (14–200 kDa)	Carl Roth (Karlsruhe, DE)

Table 3.3 Plasmids used or created in this work

Plasmid	Mutations	Mimic	Source
pRS315-*PRE2*	–	wt	W. Heinemeyer
pRS315-*pre2 Q53S*	Q53S	β5c	This work
pRS315-*pre2 S*	A46S	β5i	This work
pRS315-*pre2 C*	G48C	β5i	This work
pRS315-*pre2 SC*	A46S, G48C	β5i	This work
pRS315-*pre2 MSC*	V31 M, A46S, G48C	β5i	This work
pRS315-*pre2 SSC*	A27S, A46S, G48C	β5i	This work
pRS315-*pre2 SMSC*	A27S, V31 M, A46S, G48C	β5i	This work
pRS315-*pre2 SCT*	A46S, G48C, V127T	β5i	This work
pRS315-*pre2 SSCT*	A27S, A46S, G48C, V127T	β5i	This work
pRS315-*pre2 MSCT*	V31 M, A46S, G48C, V127T	β5i	This work
pRS315-*pre2 SMSCT*	A27S, V31 M, A46S, G48C, V127T	β5i	This work
pRS315-*pre2 SMSCRT*	A27S, V31 M, A46S, G48C, T57R, V127T	β5i	This work
pRS315-*pre2 SNSCRT*	A27S, K32 N, A46S, G48C, T57R, V127T	β5i	This work
pRS315-*pre2 SNSCRGT*	A27S, K32 N, A46S, G48C, T57R, K71G, V127T	β5i	This work
pRS315-*pre2 SNASCRGT*	A27S, K32 N, M45A, A46S, G48C, T57R, K71G, V127T	β5i	This work
pRS315-*pre2 SNVSCRGT*	A27S, K32 N, M45 V, A46S, G48C, T57R, K71G, V127T	β5i	This work
pRS315-*pre2 R*	M45R	β5i	W. Heinemeyer [6]
pRS315-*pre2 TR*	I35T, M45R	β5i	W. Heinemeyer [6]
pRS315-*pre2 SCS*	A20S, A22C, A46S	β5t	This work
pRS315-*pre2 SS*	V31S, A46S	β5t	This work
pRS315-*pre2 SCSS*	A20S, A22C, V31S, A46S	β5t	This work
pRS315-*pre2 TST*	M45T, A46S, G48T	β5t	This work
pRS315-*pre2 STST*	V31S, M45T, A46S, G48T	β5t	This work
pRS315-*pre2 SCTST*	A20S, A22C, M45T, A46S, G48T	β5t	This work
pRS315-*pre2 SCSTST*	A20S, A22C, V31S, M45T, A46S, G48T	β5t	This work

3.1.10 Instruments

Balances	
Analytical balance TE124S	Sartorius (Göttingen, DE)
Precision balance BP3100 P	Sartorius (Göttingen, DE)

Centrifuges		
Biofuge pico		Heraeus instruments (Hanau, DE)
SIGMA 4K15	rotor 11150/13220	SIGMA Laborzentrifugen

Centrifuges		
	rotor 11150/13350	(Osterode am Harz, DE)
SIGMA 6–16 K	rotor 12500	SIGMA Laborzentrifugen (Osterode am Harz, DE)
SIGMA 8 K	rotor 11805	SIGMA Laborzentrifugen (Osterode am Harz, DE)

Crystallography	
Art robbins instruments Intelli-Plates (96 well)	Dunn labortechnik (Asbach, DE)
Art robbins instruments phoenix	Dunn labortechnik (Asbach, DE)
Cooled Incubator series 3000	RUMED® Rubarth apparate (Laatzen, DE)
CrystalCap HT™ für CryoLoop™	Hampton (Aliso Viejo, US)
CrystalCap HT™ vial	Hampton (Aliso Viejo, US)
CrystalWand magnetic™	Hampton (Aliso Viejo, US)
Foam dewar	Spearlab (San Francisco, US)
Magnetic caps, pins and vials	Molecular dimensions (Newmarket, UK)
MICORLAB® STARlet	Hamilton (Reno, US)
Micro tool box	Molecular dimensions (Newmarket, UK)
Mounted CryoLoop™	Hampton (Aliso Viejo, US)
Protein crystallization screening suites	QIAGEN (Hilden, DE)
Quick combi sealer plus	HJ-Bioanalytik (Mönchengladbach, DE)
Siliconized glass cover slides	Hampton (Aliso Viejo, US)
SuperClear pregreased 24 well plate	Crystalgen (New York, US)
Vial clamp	Molecular dimensions (Newmarket, UK)
Zoom stereo microscope SZX10/KL1500LCD	Olympus (Tokio, JP)

Electrophoresis	
Chamber and tray	Appligene (Watford, UK)
Digital graphic printer UP-D897	Sony (Minato, JP)
Electrophoresis power supply EPS 600	Pharmacia biotech (Uppsala, SE)
Gel documentation system G:BOX	Syngene (Cambridge, US)
Mini PROTEAN® cell	BioRad (Hercules, US)
PowerPAc basic power supply	BioRad (Hercules, US)

Liquid chromatography	
ÄKTAprime™ plus	GE Healthcare (Chalfont St. Giles, UK)
ÄKTApurifier™	GE Healthcare (Chalfont St. Giles, UK)
CHT™Ceramic hydroxyapatite	BioRad (Hercules, US)
Phenyl sepharose 6 fast flow	GE Healthcare (Chalfont St. Giles, UK)
RESOURCE™ Q, 6 ml	GE Healthcare (Chalfont St. Giles, UK)
HiPrep™ 26/10 desalting column	GE Healthcare (Chalfont St. Giles, UK)

Additional equipments and materials	
Cary eclipse fluorescence spectrometer	Varian (Darmstadt, DE)
NanoPhotometer™ pearl	IMPLEN (München, DE)
Ultraspec10 cell density meter	Amersham bioscience (Uppsala, SE)

Additional equipments and materials	
Constant cell disruption system E1061	Constant systems (Northants, UK)
Electroporation cuvette, 2 mm	PeqLab (Erlangen, DE)
Gene pulser mit pulse controller	BioRad (Hercules, US)
Incubator	Binder (Tuttlingen, DE)
Infors HT multitron 2 cell shaker	INFORS HT (Bottmingen/Basel, CH)
inoLab® pH 720 pH-Meter	WTW (Weilheim, DE)
Laboklav 25/195	SHP Steriltechnik (Magdeburg, DE)
MR Hei-standard magentic stirrer	Heidolph (Schwabach, DE)
Techne Dri-Block DB 2A	Bibby scientific (Stone, UK)
Thermomixer comfort	Eppendorf (Hamburg, DE)
Thermocycler MyCycler™	BioRad (Hercules, US)
Vortex genie 2	Scientific Industries (New York, US)
White 96 well plate NUNC	Thermo scientific (München, DE)

3.1.11 Computer Software and Bioinformatics Tools

ApE—A plasmid editor	Wayne Davis
BOBSCRIPT	Esnouf [7]
CCP4 Software suite	www.ccp4.ac.uk [8]
ChemDraw	Perkin Elmer (Cambridge, US)
Coot	Emsley [9]
CorelDRAW X5	Corel (Ottawa, CA)
DNAman	Lynnon corporation (Quebec, CA)
EndNote X4	Adept scientific (Frankfurt, DE)
GraphPad Prism 5	GraphPad software Inc. (La Jolla, US)
MAIN	Turk [10]
Microsoft office	Microsoft (Redmond, US)
MOLSCRIPT	Kraulis [11]
PyMOL molecular graphics system	Schrödinger [12]
QuikChange® primer design tool	Agilent technologies
Sybyl	Tripos (St. Louis, US) [13]
UNICORN™ control software	GE Healthcare (Chalfont St. Giles, UK)
XDS Program package	Kabsch (Heidelberg, DE) [14]

DNA and protein sequences were obtained from the "Universal protein resource" (www.uniprot.org). The ProtParam tool (http://www.expasy.ch/tools/protparam.html) was used to calculate physical and chemical parameters such as molecular mass, pI and the extinction coefficient of protein sequences. Multiple DNA or protein sequences were aligned with the T-Coffee tool (http://www.ebi.ac.uk/Tools/t-coffee/index.html). For the calculation of protein surface areas the online tool PISA (Protein interfaces, surfaces and assemblies service PISA at European Bioinformatics Institute (http://www.ebi.ac.uk/pdbe/prot_int/pistart.html) [15]) was used.

3.2 Genetic Methods

3.2.1 *Cultivation and Long-Term Storage of Escherichia coli*

E. coli were cultivated either in liquid LB_{Amp} medium or on LB_{Amp} agar plates. Liquid cultures were inoculated from single colonies and incubated at 37 °C and 130 rpm in a shaker. Streaks on agar plates were grown overnight at 37 °C and later on stored at 4 °C in the fridge. For long-term preservation of *E. coli* clones glycerol stocks were prepared: an overnight culture was pelleted for 5 min at 3,000 rpm, cells were resuspended in LB_0 medium containing 30 % (v/v) glycerol, transferred into a cryo tube and frozen in liquid nitrogen.

3.2.2 *Cultivation and Long-Term Storage of Saccharomyces cerevisiae*

Baker's yeast was cultivated at 30 °C either on agar plates or in liquid medium. Wt and mutant yeast strains were streaked on YPD plates or on CM plates containing the appropriate selection marker composition. Agar plates were stored for 2–3 days at 30 °C and later on kept at 4 °C in the fridge. For the large scale cultivation of yeast in liquid medium 50–300 ml YPD medium were inoculated with a single colony grown on a YPD agar plate. After incubation for 12–24 h at 130 rpm this preculture was used to inoculate the main culture of 3–18 l of YPD medium in a ratio of 1:50. The main culture was grown up to 48 h at 30 °C and 130 rpm to an $OD_{600\,nm}$ of approximately 7 and subsequently harvested for 20 min at 5,000 rpm. The cells were washed in ddH_2O, transferred into 50 ml falcons, centrifuged again and finally frozen at −20 °C. For long-term storage of *S. cerevisiae* clones glycerol stocks were prepared: With a sterile toothpick cells were scraped off from an agar plate, resuspended in 1 ml of 15 % (v/v) glycerol, transferred into a cryo tube and frozen at −80 °C.

3.2.3 *Polymerase Chain Reaction*

The polymerase chain reaction (PCR) served to amplify target genes encoded on genomic DNA or plasmids [16]. Annealing temperatures were chosen 5 °C below the melting temperatures (T_m) of the used primers. The T_m was calculated according to formula 1 or by the QuikChange® Primer Design Tool from Agilent Technologies.

$$T_m = 2\,^\circ\mathrm{C} \cdot \left(B_{adenine} + B_{thymine}\right) + 4\,^\circ\mathrm{C} \cdot \left(B_{guanine} + B_{cytosine}\right)$$

Formula 1: Calculation of the melting temperature of oligonucleotides.

In general, PCR reactions were performed according to the pipetting and temperature schemes given in the Tables 3.4 and 3.5.

Table 3.4 Components of a PCR reaction using the Phusion® DNA polymerase

Compound	Final concentration	Volume [μl]
Template	0.001–0.01 ng/μl	1–5
dNTP-Mix (10 mM)	0.2 mM	2
Forward primer (10 pmol/μl)	0.5 μM	5
Reverse primer (10 pmol/μl)	0.5 μM	5
Phusion® HF buffer (5x)	1x	20
Phusion® DNA polymerase (2 U/μl)	0.02 U/μl	1
ddH_2O		Fill up to 100

Table 3.5 Temperature program for PCR reactions using the Phusion® DNA polymerase

PCR step	Temperature (°C)	Time	
Initial denaturing	95	3 min	
Denaturing	95	30 s	35 Cycles
Annealing	55	30 s	
Elongation	72	variable	
Final elongation	72	10 min	
Cooling	4	∞	

Table 3.6 Components of a PCR reaction using the Pfu Turbo DNA polymerase®

Compound	Final concentration	Volume [μl]
Template	0.2–1 ng/μl	x
dNTP-Mix (10 mM)	0.2 mM	1
Forward primer (10 pmol/μl)	2.5 ng/μl	y
Reverse primer (10 pmol/μl)	2.5 ng/μl	z
Pfu buffer (10x)	1x	5
Pfu Turbo DNA polymerase®		1
ddH_2O		Fill up to 50

Table 3.7 Temperature program for QuikChange® PCR reactions

PCR step	Temperature (°C)	Time	
Initial denaturing	95	30 s	
Denaturing	95	30 s	16 cycles
Annealing	55	1 min	
Elongation	68	7.5 min	
Cooling	4	∞	

For site-directed mutagenesis alternatively the QuikChange® PCR was used (see also Sect. 3.2.4 and Tables 3.6 and 3.7).

3.2.4 Site-Directed Mutagenesis

Mutations in the yeast proteasome gene *PRE2* encoding the subunit yβ5 were introduced either via recombinant PCR techniques or QuikChange® mutagenesis.

The first method required three PCR reactions and four oligonucleotides two of which encoding the desired mutation:

PCR 1a: Primer 5/6F and Mutation_rev Primer; elongation time 45 s

PCR 1b: Primer 5/6R and Mutation_for Primer; elongation time 45 s

PCR 2: Primer 5/6F and 5/6R with the PCR products 1a and 1b as template; elongation time 90 s.

The PCR product 2 was digested, inserted into the vector pRS315 and subsequently transformed into *E. coli*.

The QuikChange® mutagenesis [17] required two HPLC purified primers, containing the desired mutation. Usually, different amounts of template DNA were tested. After the PCR reaction 1 μl of the restriction endonucleoase *Dpn*I was added and the sample was incubated for 1.5 h at 37 °C. *Dpn*I selectively degraded only the methylated template DNA. Finally, 3 μl DNA were transformed into *E. coli* and all cells were streaked on an LB_{Amp} agar plate.

3.2.5 Agarose Gel Electrophoresis

For analytical and preparative separations of DNA 1 % (w/v) agarose gels in 1x Tris–Acetate-EDTA buffer (TAE) were used. Samples were supplemented with DNA loading dye (final concentration: 1x) and loaded on the gel. Electrophoresis was carried out for 40 min in 1x TAE buffer at 120 V. Staining of the gel in a solution of 1 mg/l ethidium bromide for 20 min enabled visualization of DNA bands under UV light of 365 nm.

DNA loading dye (10x)	Tris–HCl, pH 8.2	10 mM
	EDTA	1 mM
	Glycerol	50 % (v/v)
	Xylene cyanole	0.25 % (w/v)
	Bromphenol blue	0.25 % (w/v)
TAE buffer (50x)	Tris-Ac, pH 8.2	2 M
	EDTA	100 mM

3.2.6 Isolation of Plasmid DNA from E. coli

Isolation of plasmid DNA from *E. coli* was carried out with the peqGOLD Plasmid Miniprep Kit I or II according to the instruction manual. Purified plasmids were stored at −20 °C.

3.2.7 Purification of DNA

DNA was purified by the peqGOLD Gel extraction or the peqGOLD Cycle-Pure Kit according to the manufacturer's instruction. Purified DNA samples were stored at −20 °C.

Table 3.8 Pipetting scheme for preparative restriction digests

Compound	Final concentration	Volume [µl]
DNA		Up to 52
NEB buffer 2 (10x)	1x	7
BSA (10x)	1x	7
*Bam*HI (20 U/µl)	0.6 U/µl	2
*Hind*III (20 U/µl)	0.6 U/µl	2
H_2O		Add to 70

Table 3.9 Pipetting scheme for analytical restriction digests

Compound	Final concentration	Volume [µl]
Plasmid		2
NEB buffer 2 (10x)	1x	1
BSA (10x)	1x	1
*Bam*HI (20 U/µl)	0.5 U/µl	0.25
*Hind*III (20 U/µl)	0.5 U/µl	0.25
H_2O		5.5

3.2.8 Restriction Digest of DNA

Preparative restriction digests of a total volume of 70 µl contained 40 U per restriction endonuclease in the recommended buffer and were incubated for 2–5 h at 37 °C. Analytical cleavages were performed in a total volume of 10 µl containing the appropriate buffer and 5 U per restriction endonuclease. Analytical digests were stored for 1.5 h at 37 °C (see Tables 3.8 and 3.9).

3.2.9 Determination of DNA Concentration

DNA concentrations were determined by measuring the absorption at 260 nm with the nanophotometer (IMPLEN); an $OD_{260\ nm}$ of 1 corresponds to a concentration of 50 ng/µl DNA.

3.2.10 Ligation of DNA Fragments

Vector and insert amounts were calculated according to Formula 2.

$$m_{insert} = \frac{m_{vector} \cdot bp_{insert}}{bp_{vector}} \cdot x$$

Formula 2: Calculation of vector and insert amounts for the ligation of DNA fragments. For standard reactions m_{vector} was 50–100 ng and x = 3–5.

Table 3.10 Pipetting scheme for the ligation of DNA fragments

Compound	Final concentration	Volume [μl]
Plasmid	5–10 ng/μl	x
Insert		y
H_2O		7.5−(x + y)
T4 DNA ligation buffer (5x)	1x	2
T4 DNA ligase (1 U/μl)	0.05 U/μl	0.5

To unwind DNA fragments both plasmid and insert were incubated for 10 min at 55 °C and for 5 min on ice. Afterwards ligation buffer and T4 ligase were added (see Table 3.10). The sample was either kept at 4 °C overnight or at room temperature for 1 h.

3.2.11 Transformation of E. coli by Electroporation

40 μl of electrocompetent *E. coli* cells and 1 μl of plasmid DNA were mixed and transferred into a 2 mm electroporation cuvette. Transformation of *E. coli* was achieved by applying a high voltage pulse of approximately 5.7 ms and 2,500 V [18]. Subsequently, cells were resuspended in 1 ml of SOC medium and incubated for 1 h at 37 °C and 500 rpm shaking to recover. Different dilutions of cells were plated on agar plates containing ampicillin.

3.2.12 Transformation of S. cerevisiae

S. cerevisiae was transformed according to the LiAc method by Gietz and Woods [19]. Yeast cells were picked with a sterile toothpick and resuspended in 1 ml of sterile water. After a short spin down at 13,000 rpm for 30 s the supernatant was discarded and the pelleted cells were taken up in 1x TE/LiAc buffer. An additional step of centrifugation followed. Then, yeast cells were resuspended in x times 50 μl 1x TE/LiAc and x times 4 μl boiled single stranded salmon carrier DNA from salmon testes (Sigma-Aldrich (St. Louis, US)). The resulting cell suspension was split into x tubes and each was supplemented with 5 μl of plasmid DNA to be transformed and 300 μl of PEG3350/TE/LiAc (8:1:1). The samples were gently mixed and incubated for 30 min at 30 °C. After cells have been heated to 42 °C for 20 min, they were centrifuged for 2 min at 2,000 rpm, resuspended in 100 μl 1x TE buffer and streaked on CM ura$^-$ leu$^-$ his$^-$ plates.

TE buffer (10x)	Tris-HCl, pH 7.4	100 mM
	EDTA	10 mM
LiAc (10x)	LiAc, pH 7.5	1 M
PEG3350 (10x)	PEG3350	50 % (w/v)

3.2.13 Plasmid Shuffling

Yeast proteasome mutants were created by the plasmid shuffling procedure [20]. The haploid yeast strain YWH20a, which is chromosomally deleted in its wildtype *PRE2* gene and instead carries a wt gene copy on an *URA3* episome (pRS316), was transformed by *pre2* gene variants, encoded on the plasmid pRS315 carrying the *LEU2* selection marker (see Table 3.3).

2–3 days after the transformation single colonies were picked, streaked either on an YPD or CM leu$^-$ his$^-$ plate as patches and grown for additional 2 days at 30 °C. To select for clones that have lost the wt *PRE2* gene copy, transformants were grown on CM leu$^-$ his$^-$ medium containing 5-FOA. Optionally, after 2–3 days of incubation at 30 °C a second round of 5-FOA selection can be made. Finally, single colonies were picked from the 5-FOA containing plates and transferred onto YPD agar. Due to the indispensable need for a functional *pre2* gene for yeast survival no further selection was required.

3.2.14 DNA Isolation from S. cerevisiae

Yeast cells were scraped off from an agar plate with a toothpick and resuspended in 500 μl of ddH$_2$O. Cells were centrifuged for 30 s at 14,000 rpm. The cell pellet was taken up in 200 μl of breaking buffer. After the addition of 200 μl of glass beads (0.5 mm) and 200 μl of phenol/chloroform/isoamyl alcohol (25:24:1) solution, the samples were vortexed for 2 min. Cell debris was pelleted by centrifugation for 5 min at 14,000 rpm. 100 μl supernatant were mixed with 10 μl 3 M sodium acetate, pH 6.0 and 280 μl 100 % ethanol and subsequently stored at −80 °C for 5 min to precipitate nucleic acids. After a centrifugation step for 5 min at 14,000 rpm the supernatant was discarded and the pellet containing the nucleic acids was washed with 300 μl 70 % ethanol. Following an additional centrifugation step (5 min; 14,000 rpm) the pellet was dried at room temperature and finally dissolved in 50–100 μl TE-buffer and 0.5–1 μl RNase (10 μg/μl). 1 μl of this solution was used as a template for further PCR analysis.

Breaking buffer	Tris-HCl, pH 8.0	10 mM
	NaCl	100 mM
	EDTA	1 mM
	SDS	1 % (w/v)
	Triton X-100	2 % (v/v)

3.2.15 DNA Sequencing

DNA sequence analysis was performed according to Sanger et al. [21] by GATC Biotech AG, Konstanz. The results obtained by sequencing were compared to the corresponding entries in the Uniprot database

3.3 Protein Chemistry and Analytics

3.3.1 Purification of the Yeast 20S Proteasome

Purification of wt and mutant yCPs was carried out together with the technician Richard Feicht according to published procedures [22]. 120 g yeast cells were solubilized in approximately 150 ml of 50 mM KH_2PO_4/K_2HPO_4 buffer pH 7.5. DNAse I was added and the cells were disrupted with a French press. The cell lysate was centrifuged for 30 min at 21,000 rpm at 4 °C. The resulting supernatant was filtered and 30 % of saturated ammonium sulfate was added. Subsequently, the cell lysate was loaded on a Phenyl Sepharose™ 6 Fast Flow column pre-equilibrated with 1 M ammonium sulfate in 20 mM KH_2PO_4/K_2HPO_4 buffer pH 7.5. The yCP was eluted by applying a linear gradient from 1 M to 0 M ammonium sulfate in 4 column volumes. Collected fractions were tested for proteolytic activity using the fluorogenic substrates Suc-Leu-Leu-Val-Tyr-AMC (for wt yCP) and Cbz-Leu-Leu-Glu-AMC (for β5 mutant yCPs): 30 μl of each fraction were incubated for 1 h with 1 μl of 10 mM substrate and the resulting fluorescence was measured ($\lambda_{exc} = 360$ nm and $\lambda_{em} = 460$ nm). Active fractions were pooled and applied to a hydroxyapatite column, which has been equilibrated with 20 mM KH_2PO_4/K_2HPO_4 pH 7.5. Using a linear gradient from 20 to 500 mM KH_2PO_4/K_2HPO_4 in 20 column volumes, the yCP was eluted. Proteolytically active fractions were loaded on a Resource Q column and a sodium chloride gradient from 0 to 500 mM in 20 mM Tris–HCl pH 7.5 was run over 10 column volumes. For crystallization the buffer was exchanged for 10 mM 2-(*N*-morpholino)ethanesulfonic acid (Mes) pH 6.8 using a HiPrep™ 26/10 desalting column.

3.3.2 Purification of the Murine Immuno- and Constitutive 20S Proteasome

Immuno- and constitutive proteasome samples were kindly provided by Dr. Michael Basler and Ricarda Schwab from the group of Prof. Dr. Marcus Groettrup, University of Constance, Department of Biology/Immunology, Universitätsstr. 10, Constance, DE. The cCP was purified from livers of C57BL/6 *PSMB8* (β5i) and *PSMB10* (β2i) knockout mice according to published protocols [23]. iCP samples were prepared from livers of BALB/c mice eight days after their intravenous infection with 200 plaque-forming units of lymphocytic choriomeningitis virus strain WE (LCMV-WE) [23]. LCMV-WE selectively infects liver cells and leads to an almost complete conversion of the cCP to the iCP within eight days, when mice were sacrificed and livers removed. Purification of the iCP was carried out according to established procedures [23]. For short-term storage and subsequent crystallisation trials samples were kept at 4 °C (in 10 mM Hepes pH 7.2, 300 mM KCl, 5 mM $MgCl_2$), otherwise the protein was frozen at −80 °C.

Table 3.11 Pipetting scheme for four SDS gels

Compound	Separating gel (12 %)	Stacking gel (4 %)
Acrylamide	6 ml	1 ml
Separating gel buffer	5 ml	–
Stacking gel buffer	–	5 ml
H_2O	9 ml	4 ml
APS (10 % (v/v))	100 μl	100 μl
TEMED (1 % (v/v))	10 μl	10 μl

3.3.3 SDS Polyacrylamide Gel Electrophoresis

Discontinuous SDS-PAGE was performed according to Laemmli [24] (Table 3.11). Gels were run in 1x running buffer at 25 mA per gel until the front line has passed the bottom of the gel. SDS polyacrylamide gels were stained for 30 min with Coomassie solution and afterwards destained.

Acrylamide solution	Acrylamide	39 % (w/v)
	Bisacrylamide	1.2 % (w/v)
SDS-PAGE loading dye	Tris-HCl, pH 6.8	60 mM
	Glycerol	30 % (v/v)
	Saccharose	10 % (w/v)
	SDS	5 % (w/v)
	Bromphenol blue	0.02 % (w/v)
	β Mercaptoethanol	3 % (v/v)
Separating gel buffer	Tris-HCl, pH 8.8	1.5 M
	SDS	0.4 % (w/v)
Stacking gel buffer	Tris-HCl, pH 6.8	0.5 M
	SDS	0.4 % (w/v)
Running buffer	Tris-HCl, pH 8.2–8.3	25 mM
	Glycine	192 mM
	SDS	0.1 % (w/v)
Staining solution	Coomassie brilliant blue R250	0.05 % (w/v)
	Isopropanol	25 % (v/v)
	Glacial acetic acid	10 % (v/v)
Destaining solution	Glacial acetic acid	10 % (v/v)

3.3.4 Determination of Protein Concentration

UV/VIS spectra of protein solutions were recorded with the nanophotometer of IMPLEN. The absorption at a wavelength of 280 nm was used to calculate protein concentrations via the Lambert–Beer law. The required theoretical molar extinction coefficient was computed with the ProtParam tool (yCP: 727.3 cm·l/mmol, cCP: 685.0 cm·l/mmol, iCP: 678.8 cm·l/mmol).

3.4 Proteasome Activity Tests

3.4.1 *Overlay Assay*

Yeast strains were streaked as patches on YPD plates and incubated for two days at 30 °C. Next, cells were replica plated on a steril filter paper lying on an YPD plate and grown for 48 h at 30 °C. The filter was placed in a glass Petri dish and the cells were lysed with 10 ml of chloroform for 15 min. After having dried the filter in a plastic petri dish 10 ml of hand warm overlay solution (1 % (w/v) agar, 50 mM Tris–HCl pH 8.0, 300 μl of 10 mM Cbz-Gly-Gly-Leu-pNA substrate in DMSO) were added. Incubation for 3 h at 30 °C was followed by exposure to 10 ml of 0.1 % (w/v) sodium nitrite solution in 1 M HCl for 5 min. Next, this solution was exchanged for 10 ml of 0.5 % (w/v) ammonium sulfamate solution in 1 M HCl and after 5 min again discarded. Finally, the filter was stored in 10 ml of 0.05 % (w/v) *N*-(1-naphthyl)ethylenediamine solution in 47 % (v/v) EtOH for 10–30 min. Cells with normal ChTL activity became pink, while those with defects were less coloured.

3.4.2 *Fluorescence Activity Test and Determination of IC_{50} Values*

Proteasome activity and the inhibitory potency of ligands are routinely determined by fluorescence spectroscopy using AMC-peptides as substrates. CPs were mixed with a series of different inhibitor concentrations or DMSO as a control in 100 mM Tris-HCl, pH 7.4 and samples were stored for 45 min at room temperature to allow inhibitor binding. Next, the β5-specific substrate Suc-Leu-Leu-Val-Tyr-AMC was added to a final concentration of 200 μM. During the subsequent 60 min incubation at room temperature residual proteasome activity hydrolysed the substrate and released the AMC-fluorophor. After dilution (1:10) of the samples with 20 mM Tris-HCl, pH 7.4, which strongly slowed down the reaction, the relative fluorescence units (RFU) were measured with a Varian Cary Eclipse Fluorescence Spectrophotometer (Agilent Technologies) at $\lambda_{exc} = 360$ nm and $\lambda_{em} = 460$ nm. RFU values were normalized to the DMSO treated control, which should have retained nearly 100 % activity. The calculated residual activities were plotted against the logarithm of the applied inhibitor concentration and fitted to the equation Y = Bottom + (Top–Bottom)/(1 + 10 ^ ((LogIC50-X) * HillSlope)) (X: logarithm of inhibitor concentration; Y: Residual enzymatic activity; log(inhibitor) versus response—variable slope (four parameters), GraphPad Prism 5). The IC_{50} value, the ligand concentration that leads to 50 % inhibition of the enzymatic activity, was deduced from the fitted data.

3.5 Protein Crystallography

3.5.1 *Crystallization of the Yeast 20S Proteasome*

Purified yCP was concentrated to approximately 40 mg/ml at 5,000 rpm using an Amicon® Ultra-15 Centrifugal Filter device of 100 kDa cut-off. Crystals of the yCP were grown at 20 °C by the hanging drop vapour diffusion method in drops containing a 1:1 mixture of reservoir and protein solution. The reservoir solution (300 μl) was composed of 25 mM magnesium acetate, 100 mM Mes pH 6.8 and 9–13 % (v/v) 2-methyl-2,4-pentanediol (MPD). Crystals were cryoprotected by the addition of 5 μl of 20 mM magnesium acetate, 100 mM Mes, pH 6.8 and 30 % (v/v) MPD and immediately frozen in liquid nitrogen. Crystal soaking experiments were performed by supplementing crystal drops with 5 μl cryoprotection solution and 0.3–3.0 mM of inhibitor for 2–24 h. Afterwards crystals were supercooled in liquid nitrogen.

3.5.2 *Crystallization of the Murine 20S Proteasomes*

Protein samples were concentrated to approximately 30 mg/ml at 5,000 rpm using Millipore centrifugal devices with a cut-off of 30 kDa. Initial crystallization trials with the murine cCP and iCP were set up in 96-well sitting drop plates with a final drop size of 0.2 μl containing a protein:reservoir ratio of 1:1. Hanging drop plates were set up with final drop volumes of 1 μl. All plates were stored at 20 °C and crystals grew from MPD conditions within a few days. The iCP crystallized from a reservoir solution containing 0.2 M sodium iodide and 40 % (v/v) MPD, whereas cCP crystals preferentially grew from 0.2 M sodium formate or 0.2 M potassium acetate and 40 % (v/v) MPD. After addition of 1–5 μl of reservoir solution cCP and iCP crystals were frozen in liquid nitrogen. cCP and iCP complex structures with ONX 0914 were obtained by soaking crystals with 3 mM ONX 0914 (dissolved in DMSO) in 1–5 μl of reservoir solution for at least 8 h prior to freezing.

3.5.3 *Data Collection, Processing and Structure Determination*

Diffraction data were collected at the Swiss Light Source (SLS), Villigen, Switzerland at the beamlines X06SA (Pilatus 6 M detector) and X06DA (MAR CCD detector) at a wavelength of 1.0 Å. For all data sets cryoprotection (100 K) was used. Indexing, integration and scaling of the obtained data was performed with the program package XDS [25]. All proteasome structures were solved by

Patterson search calculations with PHASER using the maximum likelihood method [26]. For yCP:ONX 0914 complex structures the yCP apo crystal structure (PDB ID 1RYP [27]) served as a starting model and structure determination was performed as previously published [28]. Structure elucidation of the cCP and iCP was carried out by molecular replacement with the coordinates of the bovine cCP (PDB ID 1IRU [29]). For rigid body, TLS (Translation/Libration/Screw) and positional refinements REFMAC5 [30] was used. Fourfold and twofold non-crystallographic symmetry averaging was applied for the cCP and iCP in refinement and model building, respectively. Model building was performed with the interactive three-dimensional graphic programs MAIN [10] and COOT [9]. Waters were placed with Arp/Warp [8]. Ligands were built with Sybyl [13] and their topology and parameter files were created by Sketcher [8]. Graphical illustration of crystal structures was performed with the programs MOLSCRIPT [11], BOBSCRIPT [7] and PyMOL [12]. Conolly surfaces were calculated and depicted with GRASP [31]. The final coordinates were proven to have good stereochemistry according to the Ramachandran plot by using Procheck [8] as well as reasonable R_{work}, R_{free}, r.m.s.d. bond and angle values (see Tables 4.1, 4.2). The coordinates of published structures were deposited in the PDB under the following accession codes: 3UNH (iCP), 3UNE (cCP), 3UNF (iCP:ONX 0914), 3UNB (cCP:ONX 0914), 3UN8 (yCP:ONX 0914_ep), 3UN4 (yCP:ONX 0914_mo).

References

1. W.O. Bullock, J.M. Fernandez, J.M. Short, XL1-Blue: a high efficiency plasmid transforming recA *Escherichia coli* strain with beta-galactosidase selection. Biotechniques **5**, 376–379 (1987)
2. D. Hanahan, Studies on transformation of Escherichia coli with plasmids. J. Mol. Biol. **166**, 557–580 (1983)
3. W. Heinemeyer, A. Gruhler, V. Mohrle, Y. Mahe, D.H. Wolf, PRE2, highly homologous to the human major histocompatibility complex-linked RING10 gene, codes for a yeast proteasome subunit necessary for chrymotryptic activity and degradation of ubiquitinated proteins. J. Biol. Chem. **268**, 5115–5120 (1993)
4. W. Heinemeyer, J.A. Kleinschmidt, J. Saidowsky, C. Escher, D.H. Wolf, Proteinase yscE, the yeast proteasome/multicatalytic-multifunctional proteinase: mutants unravel its function in stress induced proteolysis and uncover its necessity for cell survival. EMBO J. **10**, 555–562 (1991)
5. W. Heinemeyer, M. Fischer, T. Krimmer, U. Stachon, D.H. Wolf, The active sites of the eukaryotic 20 S proteasome and their involvement in subunit precursor processing. J. Biol. Chem. **272**, 25200–25209 (1997)
6. R.J.C. Estiveira,The active subunits of the 20S proteasome in *Saccharomyces cerevisiae*—Mutational analysis of their specificities and a C-terminal extension, PhD thesis, Universität Stuttgart (2008)
7. R.M. Esnouf, An extensively modified version of MolScript that includes greatly enhanced coloring capabilities. J. Mol. Graph. Model. **15**(132–134), 112–133 (1997)
8. Collaborative Computational Project, The CCP4 suite: programs for protein crystallography. Acta Crystallogr Sect. D - Biol. Crystallogr. **50**, 760–763 (1994)
9. P. Emsley, B. Lohkamp, W.G. Scott, K. Cowtan, Features and development of coot, acta crystallogr. Sect. D - Biol. Crystallogr. **66**, 486–501 (2010)

10. D. Turk, Improvement of a programme for molecular graphics and manipulation of electron densities and its application for protein structure determination, PhD thesis, *Technische Universität München* (1992)
11. P.J. Kraulis, *MOLSCRIPT*: a program to produce both detailed and schematic plots of protein structures. J. Appl. Cryst. **24**, 946–950 (1991)
12. W.L. DeLano, *The PyMOL Molecular Graphics System* (DeLano Scientific, San Carlos, 2002)
13. SYBYL 8.0 Tripos International, 1699 South Hanley Rd., St. Louis, Missouri, 63144, USA
14. W. Kabsch, Xds. Acta Crystallogr Sect. D - Biol. Crystallogr. **66**, 125–132 (2010)
15. E. Krissinel, K. Henrick, Detection of protein assemblies in crystals. CompLife **2005**(3695), 163–174 (2005)
16. K. Mullis, F. Faloona, S. Scharf, R. Saiki, G. Horn, H. Erlich, Specific enzymatic amplification of DNA in vitro: the polymerase chain reaction. Cold Spring Harb. Symp. Quant. Biol. **51**(Pt 1), 263–273 (1986)
17. C. Papworth, J.C. Bauer, J. Braman, D.A. Wright, Site-directed mutagenesis in one day with >80 % efficiency. Strategies **9**, 3–4 (1996)
18. W.J. Dower, J.F. Miller, C.W. Ragsdale, High efficiency transformation of *E. coli* by high voltage electroporation. Nucleic Acids Res. **16**, 6127–6145 (1988)
19. R.D. Gietz, R.A. Woods, Transformation of yeast by lithium acetate/single-stranded carrier DNA/polyethylene glycol method. Methods Enzymol. **350**, 87–96 (2002)
20. R.S. Sikorski, J.D. Boeke, In vitro mutagenesis and plasmid shuffling from cloned gene to mutant yeast. Academic Press Inc, San Diego 194 302–318 (1991)
21. F. Sanger, S. Nicklen, A.R. Coulson, DNA sequencing with chain-terminating inhibitors. Proc. Natl. Acad. Sci. USA **74**, 5463–5467 (1977)
22. N. Gallastegui, M. Groll, Analysing properties of proteasome inhibitors using kinetic and X-ray crystallographic studies. Methods Mol. Biol. **832**, 373–390 (2012)
23. G. Schmidtke, S. Emch, M. Groettrup, H.G. Holzhutter, Evidence for the existence of a non-catalytic modifier site of peptide hydrolysis by the 20S proteasome. J. Biol. Chem. **275**, 22056–22063 (2000)
24. U.K. Laemmli, Cleavage of structural proteins during the assembly of the head of bacteriophage T4. Nature **227**, 680–685 (1970)
25. W. Kabsch, Automatic processing of rotation diffraction data from crystals of initially unknown symmetry and cell constants. J. Appl. Cryst. **26**, 795–800 (1993)
26. A.J. McCoy, R.W. Grosse-Kunstleve, P.D. Adams, M.D. Winn, L.C. Storoni, R.J. Read, Phaser crystallographic software. J. Appl. Cryst. **40**, 658–674 (2007)
27. M. Groll, L. Ditzel, J. Löwe, D. Stock, M. Bochtler, H.D. Bartunik, R. Huber, Structure of20S proteasome from yeast at 2.4 Å resolution. Nature **386**, 463–471 (1997)
28. M. Groll, R. Huber, Purification, crystallization, and X-ray analysis of the yeast 20S proteasome. Methods Enzymol. **398**, 329–336 (2005)
29. M. Unno, T. Mizushima, Y. Morimoto, Y. Tomisugi, K. Tanaka, N. Yasuoka, T. Tsukihara The structure of the mammalian 20S proteasome at 2.75 Å resolution. Structure 10 609-618 (2002)
30. A.A. Vagin, R.A. Steiner, A.A. Lebedev, L. Potterton, S. McNicholas, F. Long, G.N. Murshudov, REFMAC5 dictionary: organization of prior chemical knowledge and guidelines for its use. Acta Crystallogr Sect. D - Biol. Crystallogr. **60**, 2184–2195 (2004)
31. A. Nicholls, K.A. Sharp, B. Honig, Protein folding and association: insights from the interfacial and thermodynamic properties of hydrocarbons. Proteins **11**, 281–296 (1991)

Chapter 4
Results

4.1 Sequence Alignments

The quaternary and tertiary structure of the 20S proteasome is strictly conserved from archaea to mammals and it is assumed that its two types of subunits, termed α and β, evolved from a common ancestor protein [1–5]. In this regard the catalytically active proteasome subunits of murine 20S proteasome types display high sequence identities to each other: β1c/β1i: 63.3 %, β2c/β2i: 58.9 %, β5c/β5i: 72.4 %, β5c/β5t: 54.4 % β5i/β5t: 50.5 % (Fig. 4.1). Although sequence alignments are indicative of an overall conserved fold for all proteolytically active subunits, defined differences in their substrate binding channels give rise to different cleavage preferences in the cCP, iCP and tCP [2, 6]. In particular, hydrophobic amino acids in the unprimed substrate binding pockets of subunit β1i have been predicted to attenuate the CL activity and to increase the ChTL activity [2]. In contrast, the sequence analysis of the subunits β2c/i and β5c/i provided no hints on changes in the chemical properties of their substrate binding channels and on profound differences in their substrate specificities (Fig. 4.1). However, for the vertebrate-specific subunit β5t, which is exclusively expressed in cTECs, the primary sequence of its substrate binding channel is strongly altered. In particular, the polarity of the binding pockets appears to be dramatically increased, thereby leading to a markedly reduced ChTL activity and to a different peptide product pattern of the tCP [7] (Fig. 4.1c; see also Sect. 4.4.3).

4.2 X-Ray Structures of the Mouse 20S Immuno- and Constitutive Proteasome

4.2.1 Crystallization and Structure Determination

Infection of BALB/c mice with LCMV-WE leads to an efficient (95 %) conversion of cCPs to iCPs in the murine livers within eight days. The synthesized iCPs were

E. M. Huber, *Structural and Functional Characterization of the Immunoproteasome*, Springer Theses, DOI: 10.1007/978-3-319-01556-9_4,

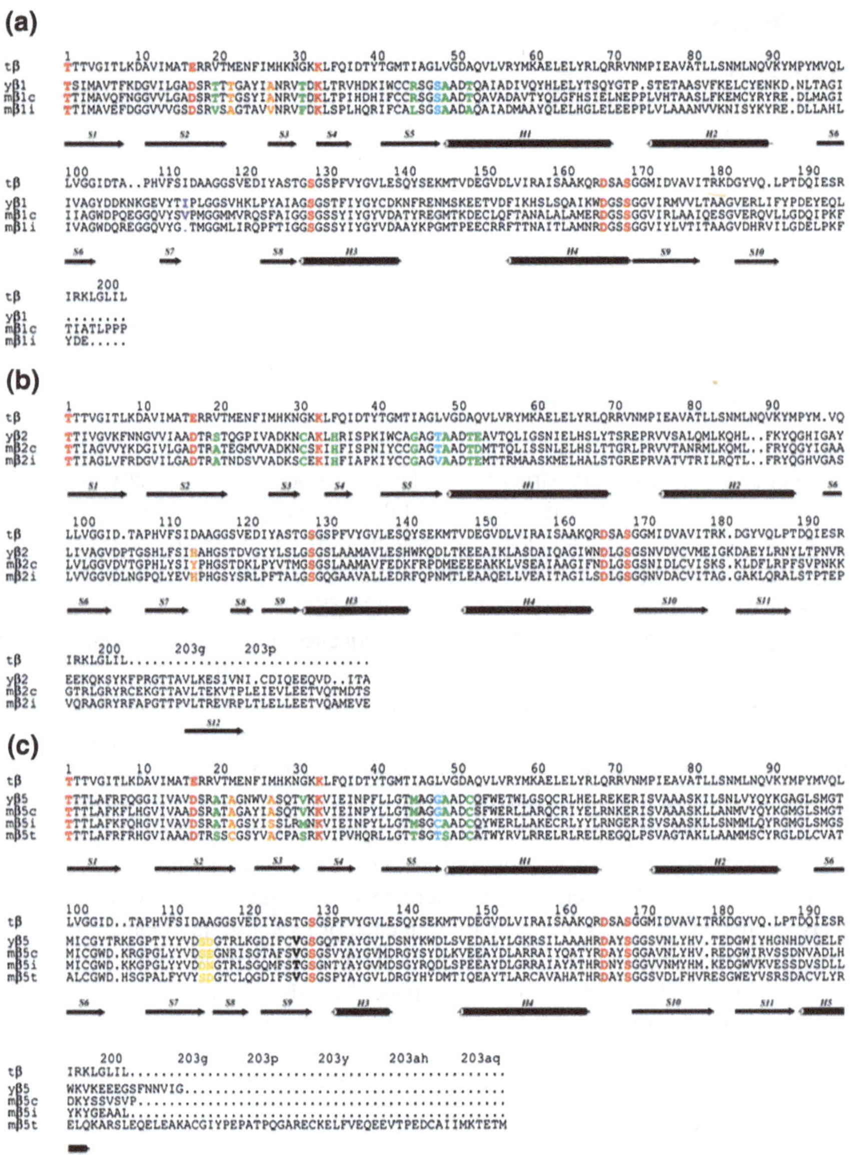

Fig. 4.1 Sequence alignments of active β subunits from archaeal (t), yeast (y) and murine (m) 20S proteasomes. Protein sequences for (**a**) β1, (**b**) β2 and (**c**) β5 subunits are given in the one-letter code. Amino acids are numbered according to the β subunit of the 20S proteasome from the archaeon *T. acidophilum* (t); insertions are indicated by lower-case letters. Helices and sheets are depicted for the i subunits. Residues that are crucial for the catalytic activity are marked in *red*. The substrate specificity pocket forming amino acids are coloured in *green* (S1), *blue* (S2), *orange* (S3) and *yellow* (S'), respectively. Residue 113 is absent in all β1i subunits (*purple*). Amino acids that are suggested to influence the active site architecture and proteolytic activity of subunit β5 are highlighted in *black*. Adapted from Huber et al. 2012 [5]

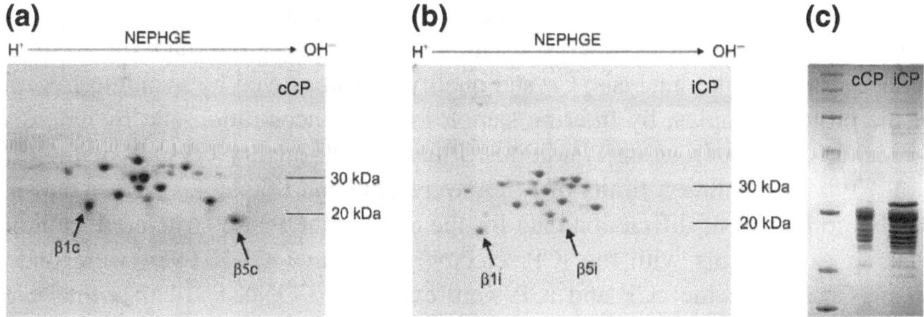

Fig. 4.2 Coomassie-stained gels of purified murine cCP and iCP samples used for crystallization trials. 2D non-equilibrium pH gradient gel electrophoresis (NEPHGE)/SDS-PAGE gels of (**a**) 80 μg of purified cCP isolated from livers of β2i $^{-/-}$ β5i $^{-/-}$ gene targeted mice and (**b**) 80 μg of iCP from livers of LCMV-WE infected BALB/c mice. Proteasome subunits were assigned according to Groettrup et al., 1996 [9] (**c**) One-dimensional SDS-PAGE analysis of cCP and iCP samples. Adapted from Huber et al. 2012 [5]

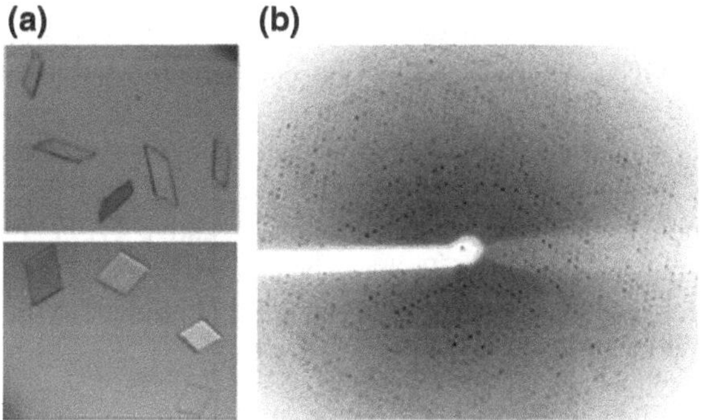

Fig. 4.3 Analysis of murine cCP and iCP crystals for X-ray diffraction. (**a**) Murine cCP and iCP crystals; (**b**) Diffraction pattern of an iCP crystal

purified according to published procedures [8]. The same purification procedure was applied to livers of uninfected gene targeted β2i$^{-/-}$ and β5i$^{-/-}$ C57BL/6 mice and yielded pure cCP samples (Fig. 4.2).

In a joint collaboration with the group of Prof. Dr. Marcus Groettrup of the University of Constance, Dr. Michael Basler and Ricarda Schwab prepared appropriate amounts of murine iCP and cCP for crystallographic analysis. Samples were proven for their purity by 2D-PAGE (Fig. 4.2), concentrated to approximately 30 mg/ml and subsequently subjected to initial sitting drop vapour diffusion crystallization trials. A sparse matrix screen of about 400 conditions was tested with a protein:reservoir ratio of 1:1 and a final drop volume of 0.2 μl. Both CPs crystallized within a few days from various reservoir solutions containing 40 % (v/v) MPD and different salts (Fig. 4.3a). The obtained crystals were exposed to synchrotron X-ray radiation and their diffraction patterns were analysed (Fig. 4.3b).

Albeit their small size, crystals diffracted to a resolution of approximately 6 Å. However, due to disordered crystal lattices most diffractions patterns were not suitable for structure analysis. Crystal quality was improved by avoiding freezing of the protein samples, by filtering samples upon concentration and by extensive screening of crystallization conditions. Finally, high-quality crystals of the cCP and iCP with resolution limits of 3.2 Å were obtained (Table 4.1). After the successful collection of diffraction data for the cCP and iCP apo-structures, inhibitor soaking experiments with the iCP-selective compound ONX 0914 were undertaken. Crystals of the cCP and iCP were exposed to ONX 0914 in a final concentration of 3 mM for at least 8 h. Diffraction data of both cCP:ONX 0914 and iCP:ONX 0914 crystals were collected to a resolution limit of 2.9 Å (Table 4.1) and data processing was performed according to Sect. 3.5.3.

The X-ray data revealed for both cCP and iCP crystals space group $P2_1$, but different unit cell parameters. Calculation of the Matthews coefficient indicated a solvent content of about 50 %, assuming that the asymmetric unit of iCP crystals contained only one CP, whereas the asymmetric unit of cCP crystals was built up of two 20S proteasomes. The same applied to both ligand complex structures (Table 4.1). All four crystal structures could be finalized with R_{free} values below 27.5 % and root-mean-square deviations for bond lengths and bond angles of less than 0.005 Å and 0.93°.

All proteasome subunits were well defined in the $2F_O$-F_C electron density maps and differences in the primary sequences of catalytically active c and i subunits were unambiguously depicted by positive and negative F_O-F_C maps (Fig. 4.4), indicating that pure iCPs and cCPs have been crystallized. Amino acid residue numbers were allocated according to the sequence alignment to the proteasomal β subunit from *T. acidophilum* (Fig. 4.1).

4.2.2 Subunit Architecture of the cCP and iCP

The overall topologies of the cCP and iCP are identical except for the exchange of the catalytically active c subunits β1c, β2c and β5c by their i homologues β1i, β2i and β5i. Likewise the yCP crystal structure, the N-termini of the inactive α subunits of the cCP and iCP close the entrance to the interior of the CPs. Superposition of the α rings from cCP, iCP and yCP illustrates their high structural similarity (r.m.s.d. C_α atoms α ring <0.59 Å; Fig. 4.5) and suggests that the gate opening mechanism [10, 11] is identical for all kind of CPs.

Subunits incorporated into both the cCP and iCP, i.e. α subunits and inactive β subunits are unchanged as proven by an r.m.s.d C_α atoms of <0.35 Å for their main chain tracings. Despite differences in their amino acid sequence (Fig. 4.1) the exchangeable β subunits adopt identical folds (r.m.s.d. C_α atoms <0.72 Å) (Fig. 4.6). Remarkably, also the C-terminal loop of the subunits yβ2 and β2c, which embraces its neighbouring subunit β3 and which is essential for CP assembly [12], is structurally conserved in subunit β2i.

Table 4.1 X-ray data collection and refinement statistics of murine iCP and cCP structures

	m_iCP	m_iCP: ONX 0914	m_cCP	m_cCP: ONX 0914
Crystal parameters				
Space group	$P2_1$	$P2_1$	$P2_1$	$P2_1$
Cell constants	a = 118.3 Å	a = 117.3 Å	a = 171.0 Å	a = 171.7 Å
	b = 205.2 Å	b = 194.6 Å	b = 201.3 Å	b = 198.6 Å
	c = 161.9 Å	c = 157.7 Å	c = 226.0 Å	c = 226.8 Å
	β = 105.7°	β = 107.1°	β = 108.1°	β = 106.6°
CPs/AU[a]	1	1	2	2
Data collection				
Beam line	X06DA, SLS	X06DA, SLS	X06SA, SLS	X06SA, SLS
Wavelength (Å)	1.0	1.0	1.0	1.0
Resolution range (Å)[b]	30-3.2	30-2.9	30-3.2	30-2.9
	(3.3-3.2)	(3.0-2.9)	(3.3-3.2)	(3.0-2.9)
No. observations	356417	392915	807596	723529
No. unique reflections[c]	121329	145087	235050	308534
Completeness (%)[b]	99.2 (99.4)	97.1 (89.8)	99.5 (99.7)	96.0 (96.7)
R_{merge} (%)[b, d]	10.4 (51.5)	11.6 (54.7)	7.6 (59.7)	8.1 (48.1)
I/σ (I)[b]	10.4 (2.3)	8.0 (2.0)	13.9 (2.4)	9.4 (2.0)
Refinement (REFMAC5)				
Resolution range (Å)	15-3.2	15-2.9	15-3.2	15-2.9
No. refl. working set	115261	137832	223279	293487
No. refl. test set	5763	6892	11163	14674
No. non hydrogen	48502	48760	97320	97172
No. of ligand atoms	–	294	–	588
Solvent (H_2O, K^+, Cl^-, J^-)	583	1045	640	1480

(continued)

Table 4.1 (continued)

	m_iCP	m_iCP: ONX 0914	m_cCP	m_cCP: ONX 0914
R_{work}/R_{free} (%)[e]	23.9/25.4	23.5/27.5	22.2/24.6	22.9/27.2
r.m.s.d. bond (Å)/(°)[f]	0.004/0.851	0.005/0.926	0.004/0.753	0.005/0.885
Average B-factor (Å^2)	79.2	61.7	78.7	67.8
Ramachandran Plot (%)[g]	97.2/2.6/0.2	96.7/3.0/0.3	95.4/4.1/0.5	95.9/3.6/0.5

[a] Asymmetric unit

[b] The values in parentheses of resolution range, completeness, R_{merge} and I/σ (I) correspond to the last resolution shell

[c] Friedel pairs were treated as identical reflections

[d] $R_{merge}(I) = \Sigma_{hkl}\Sigma_j \mid [I(hkl)_j — I(hkl)] \mid \Sigma_{hkl} I_{hkl}$, where $I(hkl)_j$ is the jth measurement of the intensity of reflection hkl and $< I(hkl) >$ is the average intensity

[e] $R = \Sigma_{hkl} \mid |F_{obs}| — |F_{calc}| \mid / \Sigma_{hkl} |F_{obs}|$, where R_{free} is calculated for a randomly chosen 5 % of reflections, which were not used for structure refinement, and R_{work} is calculated for the remaining reflections

[f] Deviations from ideal bond lengths/angles

[g] Number of residues in favoured, allowed or outlier region

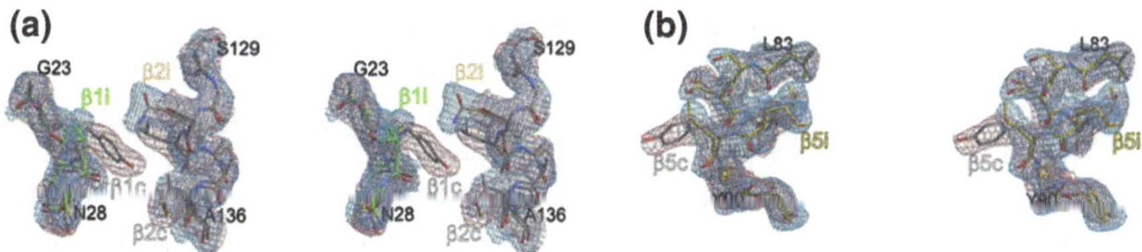

Fig. 4.4 Stereo representation of the electron density maps of the murine cCP and iCP. For the calculation of the $2F_O$-F_C electron density maps (iCP: *blue* mesh; cCP: *red* mesh), which are contoured to 0.8 σ, the phases of the displayed amino acids have been omitted. The electron density sections clearly depict explicit differences in the amino acid sequences of c and i subunits: (**a**) Tyr25 of subunit β1c corresponds to Ala25 in subunit β1i; and Ser131, Leu132 as well as Met135 of subunit β2c are substituted by Gln, Gly and Val in subunit β2i, respectively. (**b**) Tyr88 of subunit β5c is exchanged for Leu in subunit β5i. Adapted from Huber et al. 2012 [5]

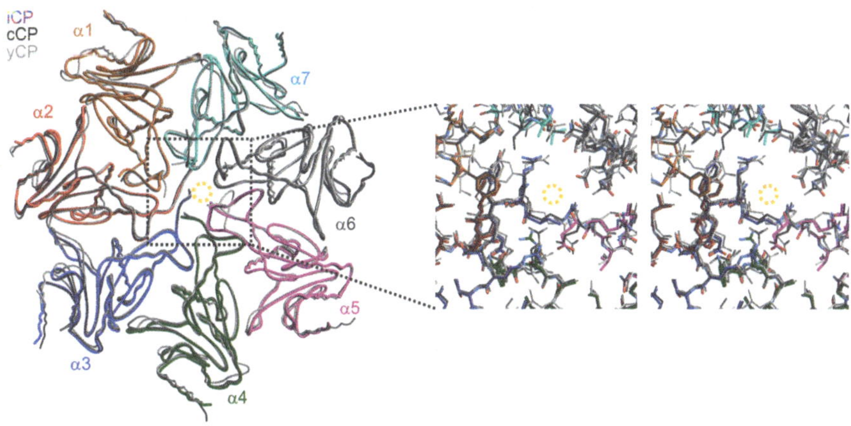

Fig. 4.5 Structural superposition of the α ring of yCP, cCP and iCP. The N-termini of the α subunits, in particular of α2, α3 and α4, from yCP (*grey*), cCP (*black*) and iCP (*multi-coloured*) prevent access into the proteolytic chamber. A stereoscopic view of the centre of the α ring (marked by a *yellow circle*) is provided. Adapted from Huber et al. 2012 [5]

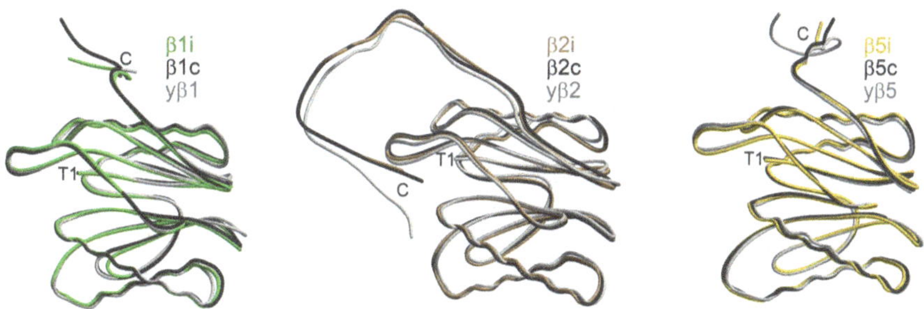

Fig. 4.6 Main chain tracings of the proteolytically active proteasome subunits. The $C_α$ backbones of the subunits β1i (*green*), β2i (*brown*) and β5i (*yellow*) are superimposed to their constitutive counterparts from mouse (*black*) and yeast (*grey*). The nucleophilic N-terminal Thr1 and the C-terminus are indicated. Adapted from Huber et al. 2012 [5]

The majority of the amino acid exchanges between the proteolytically active c and i subunits are located on the subunit surfaces and only a minority is conserved among species (β1i: 13.6 %, β2i: 5.1 %, β5i: 9.8 %). As revealed by PISA analysis, the number of interaction sites between adjacent subunits of the cCP and the iCP differ (Table A.1), but the impact on the stability or half-life of the CPs cannot be predicted from the structural data. However, the subunit interaction surfaces of the proteolytically active subunits and their adjacent neighbours enable the formation of mixed proteasome species, which were previously suggested to be physiologically relevant [13, 14].

4.2.3 Structural Analysis of the Substrate Binding Channels

The substrate and inhibitor specificities of the proteasomal active sites are predominantly determined by the enthalpic interactions of ligands with defined pockets of the primed and unprimed substrate binding channels surrounding the nucleophilic Thr1. The following section summarizes the similarities and alterations in the primed and unprimed sites of proteolytically active c and i subunits.

The primed substrate binding pockets (S') of the subunits β2i and β2c differ by several amino acid exchanges. However, their impact on protein hydrolysis remains enigmatic. In the primed site of subunit β1i one amino acid is deleted in the loop segment 113–124 compared to subunit β1c. This loop shortening is a hallmark of all β1i subunit sequences known to date and might affect substrate preferences. Comparison of the primed pockets in the subunits β5c and β5i depicts the amino acid substitutions S115D and E116N. Although these changes might influence substrate affinities, they are not conserved between species. Using X-ray crystallography most inhibitory compounds have been shown to target the unprimed substrate binding channels [15–17] and thus, the S1, S2 and S3 pockets are far better examined than the primed ones. The unprimed sites of the subunits β2c and β2i display high similarity to each other, except for the conserved substitutions T48V and D53E. While T48V might alter the specificity for P2 residues, D53E is assumed to change neither the chemical environment nor substrate preference (Fig. 4.7). Interestingly, Glu53 and Thr48 are characteristic of the yeast subunit yβ2 (Fig. 4.1).

By contrast, the β1c and β1i subunits strongly differ in their amino acid lining in the unprimed substrate binding channel. The polar active site surrounding of subunit β1c is replaced by a more hydrophobic one in subunit β1i. In particular the amino acid substitutions T20V, T31F, R45L and T52A decrease the polarity and the size of the β1i S1 pocket (Fig. 4.8). Hence, the β1i active site preferentially cleaves proteins C-terminally of small hydrophobic and branched amino acids such as Leu, Ile or Val and significantly enhances the production of high-affinity MHC I ligands. The structural features of subunit β1i are in full agreement with the reported β1i-selective fluorogenic substrate Ac-Pro-Ala-Leu-AMC [18]. In addition to the changes in the S1 pocket, the S3 pocket is characterized by the amino acid substitutions T22A (β1i) and A27V (β1i) as well as Y114H in the neighbouring subunit β2i. These differences lead to a more size-restricted and more polar S3 pocket in the iCP (Fig. 4.8).

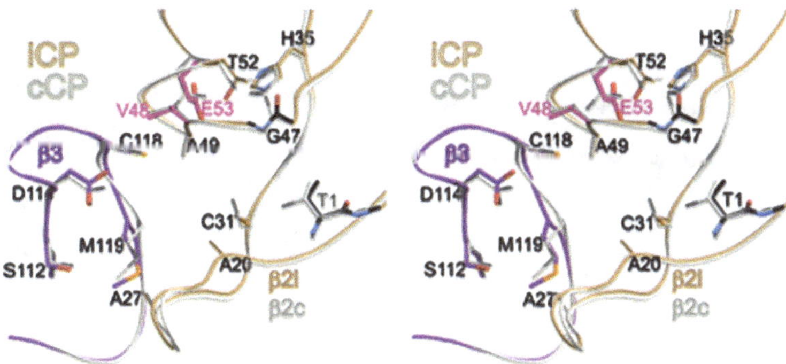

Fig. 4.7 Superposition of the unprimed β2c and β2i substrate binding channels (*Stereo view*). Amino acid numbers are given for subunit β2i and residues characteristic of β2i are highlighted in *magenta*. Adapted from Huber et al. 2012 [5]

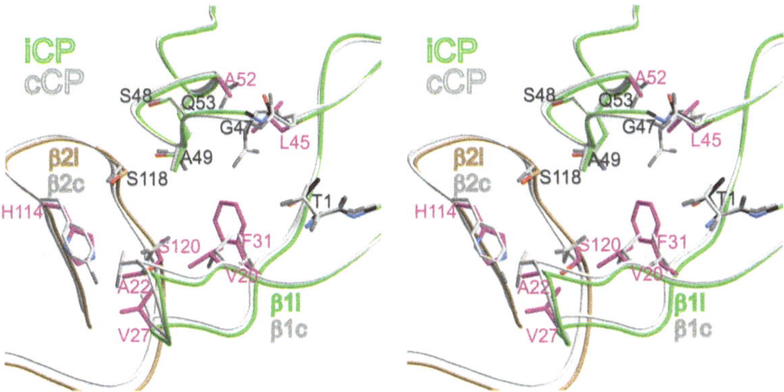

Fig. 4.8 Superposition of the unprimed β1c and β1i substrate binding channels (*Stereo view*). Amino acid numbers are given for subunit β1i and residues characteristic of β1i are highlighted in *magenta*. Adapted from Huber et al. 2012 [5]

The hydrophobic amino acid lining of the S1 pocket in subunit β5c is preserved in subunit β5i with respect to the amino acids Ala20, Met45, Ala49 and Cys52 (Fig. 4.9). The amino acid side chain 31 is also hydrophobic but variable in length (e.g. Met in mice and Val in humans). Thus, both subunits exert overlapping substrate specificities and hydrolyse proteins after apolar residues. Interestingly, all known β5i subunits possess a shallow S2 pocket formed by either Cys48 or Ser48 (Fig. 4.9). The incorporation of Ser27 in the S3 pockets of murine and human β5i subunits reduces their size but concomitantly increases their polarity.

A striking feature of the β5i subunit is the increased size of its S1 pocket compared to subunit β5c. This is mostly due to distinct conformations of Met45 and Ile35 in subunit β5i as well as a shift of the protein backbone including the residues 36–76 compared to subunit β5c (Fig. 4.9). In this regard, the amino acid exchange S53Q between subunit β5c and β5i might have a pivotal function. In subunit β5i the aliphatic part of the Gln53 side chain can stabilize Met45, while in subunit β5c Ser53

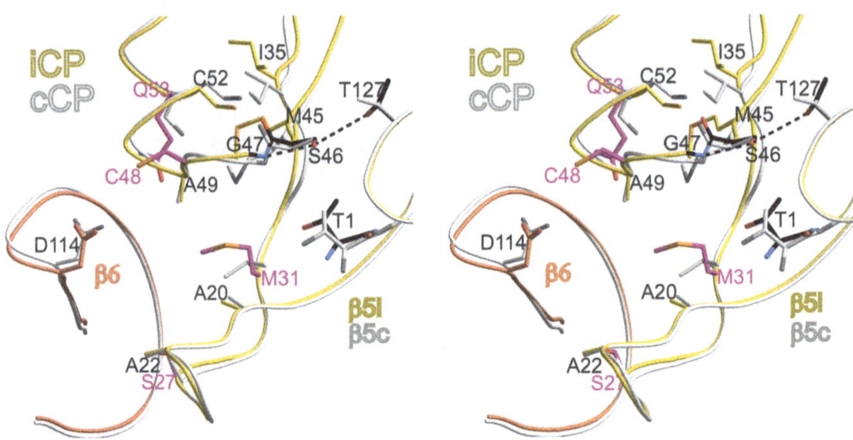

Fig. 4.9 Superposition of the unprimed β5c and β5i substrate binding channels (*Stereo view*). Amino acid numbers are given for subunit β5i and residues characteristic of the β5i substrate binding pockets are highlighted in *magenta*. Being hydrogen-bridged to the oxyanion hole Gly47NH and to Thr127 (*black dashed lines*), Ser46 is supposed to influence the catalytic activity of subunit β5i. Note that the side chain conformations of Ile35 and Met45 vary in the subunits β5c and β5i. Adapted from Huber et al. 2012 [5]

cannot provide these favourable interactions and thus, the S1 pocket is significantly smaller than that of its i counterpart (see also Sect. 4.4.1). The differently-sized S1 pockets are supposed to give rise to varying substrate specificities. Although both subunits β5c and β5i accommodate hydrophobic amino acids in their S1 pockets, the β5i subunit prefers more spacious ones than subunit β5c [18] (see also Sect. 4.3.3).

Moreover, the active site Thr1 of subunit β5i is surrounded by a unique hydrophilicity that results from the amino acid exchanges A46S and V127T and that is not observed in any other proteasome subunit. Ser46O$^{\gamma}$ is in hydrogen bonding distance to T127O$^{\gamma}$ and the oxyanion hole Gly47NH (distances 2.9–3.4 Å; Fig. 4.9). This unique hydrogen bond network might stabilize the tetrahedral transition state during catalysis. Additionally, it could kinetically favour the β5i active site, as the increased polar environment is supposed to attract water molecules and thereby to enhance peptide bond hydrolysis [5].

4.3 Investigations on the β5i-Selective Proteasome Inhibitor ONX 0914

4.3.1 Inhibition of 20S Proteasomes by ONX 0914

The α',β'epoxyketone inhibitor ONX 0914 has been shown to selectively inhibit the β5i subunit of the murine and human iCP and to be therapeutically active in inflammatory disorders and autoimmune diseases [19–22]. To verify these results and to further assess the purities of the murine cCP and iCP preparations,

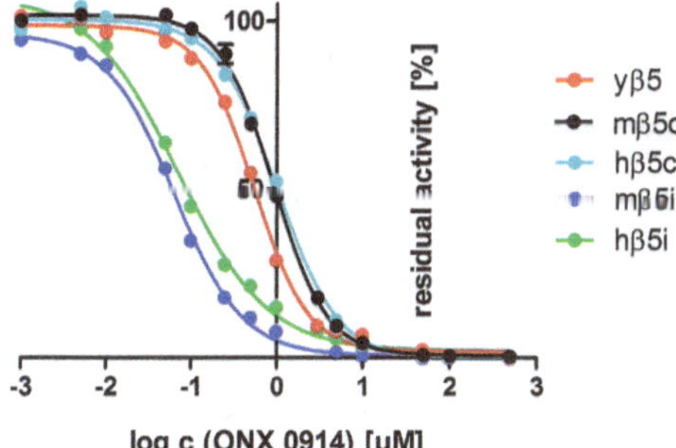

β5-subunit	*IC_{50} [μM]*
yβ5	0.551 ± 0.035
mβ5c	0.917 ± 0.054
hβ5c	1.036 ± 0.054
mβ5i	0.065 ± 0.005
hβ5i	0.073 ± 0.011

Fig. 4.10 Selectivity of ONX 0914 for the β5i subunit of the iCP. After the exposure of yCP, cCP and iCP to varying concentrations of ONX 0914 (0.001-500 μM) the residual ChTL activity was determined. Data from three experiments were normalized to DMSO treated controls and averaged. Standard deviations are indicated. IC_{50} values were deduced from fitted data. Adapted from Huber et al. 2012 [5]

IC_{50} values of ONX 0914 were determined for the ChTL activity of the yCP, the murine cCP and iCP as well as the human cCP and iCP (Fig. 4.10). The activities of CPs were measured in the presence of varying concentrations of ONX 0914, using the fluorogenic substrate Suc-Leu-Leu-Val-Tyr-AMC. In agreement with previous studies, the β5c subunits from mouse and human cCPs were far less inhibited than their i counterparts (Fig. 4.10). Interestingly, however, ONX 0914 slightly favoured subunit yβ5 over β5c.

4.3.2 Proteasome Core Particles in Complex with ONX 0914

Structural explanations for the β5i selectivity of ONX 0914 could only be unravelled by analysing and comparing ligand binding to each active site of the cCP/yCP and iCP. Inhibitor soaking experiments with a final concentration of ONX 0914 of 3 mM blocked all proteolytic centres—not only the preferred subunit β5i. $2F_O$-F_C electron density maps depicted that ONX 0914 was covalently bound to all active sites of cCP, iCP and yCP (Fig. 4.11) and additionally proved that these were catalytically active in the crystals. Despite suggestions that subunit β7 might also exert hydrolytic activity in the ligand-free bovine cCP crystal structure [4], ONX 0914 was not bound to this subunit.

Likewise other peptidic CP inhibitors, the C-terminal dipeptide of ONX 0914 forms an antiparallel β sheet in the unprimed substrate binding channels of all active sites and is well stabilized by favourable interactions with the surrounding protein residues. In contrast, the N-terminal morpholine moiety of ONX 0914 is not engaged in any contact with the substrate binding channels and thus, can adopt different conformations. Unique to all β2 and β5 subunits is the additional stabilization of ONX 0914 via a hydrogen bond between its N-terminal peptide bond and Asp114 from the neighbouring subunits β3 and β6, respectively. Furthermore, structural

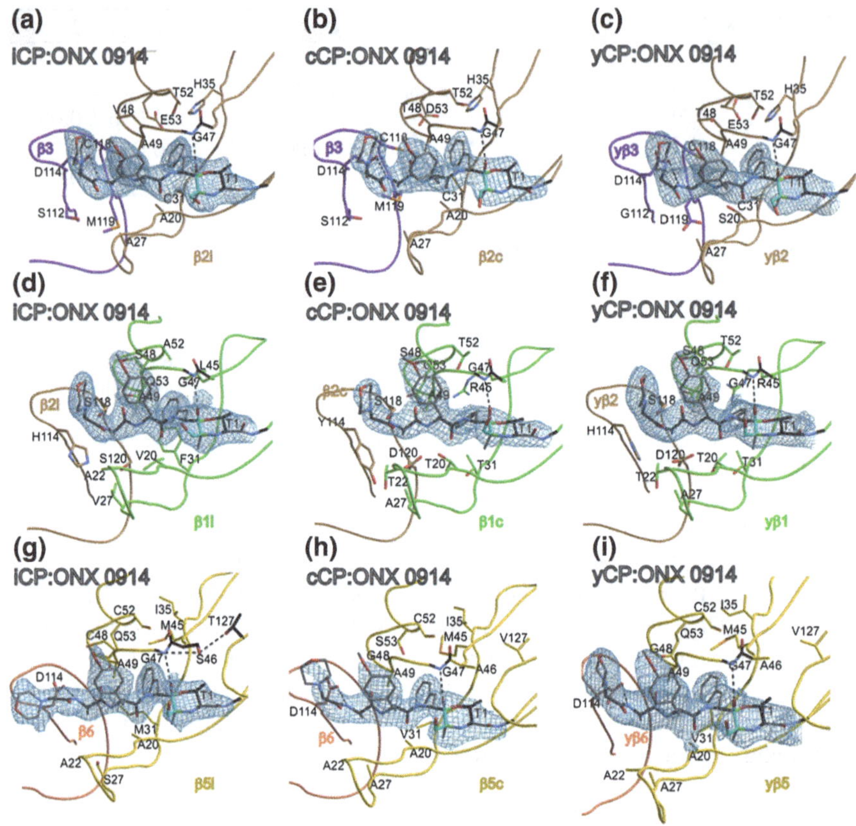

Fig. 4.11 ONX 0914 bound to the proteolytically active sites of yCP, cCP and iCP. The $2F_O$-F_C electron density maps, contoured to 1 σ, show full occupancy for ONX 0914 at all active centres of yCP, cCP and iCP. ONX 0914 and residue Thr1 have been omitted for phasing. Hydrogen bonds are marked by *black dashed lines*. Adapted from Huber et al. 2012 [5]

comparison of all ONX 0914 molecules bound to the yCP, cCP and iCP proves a similar mechanism of inhibitor binding for all active sites (Fig. 4.12).

4.3.3 Molecular Basis for the Subunit Selectivity of ONX 0914

Comparison of active proteasome subunits in the presence and absence of ONX 0914 visualized the structural features that cause the observed β5i selectivity of this compound. Owing to Ala20, Ala27, Cys31 and Gly45 the β2 subunits possess large S1 pockets that can easily accommodate ONX 0914 without any structural changes. However, as the protein surrounding provides only few stabilizing interactions, ligand binding to the β2c and β2i active sites is energetically disadvantaged compared to subunit β5i (Fig. 4.13a-c).

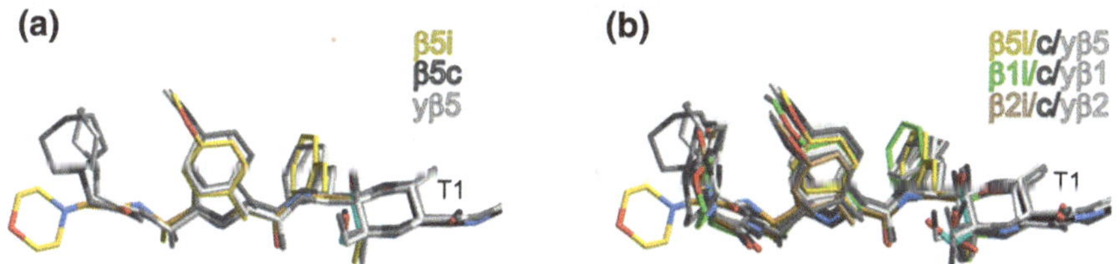

Fig. 4.12 Superposition of ONX 0914 molecules bound to the yCP, cCP and iCP. (**a**) The binding mechanism of ONX 0914 to Thr1 is identical for all β5 active sites (**b**) for all proteasome subunits. The peptidic ligand adopts an anti-parallel β sheet in each substrate binding channel, thereby mimicking natural protein substrates

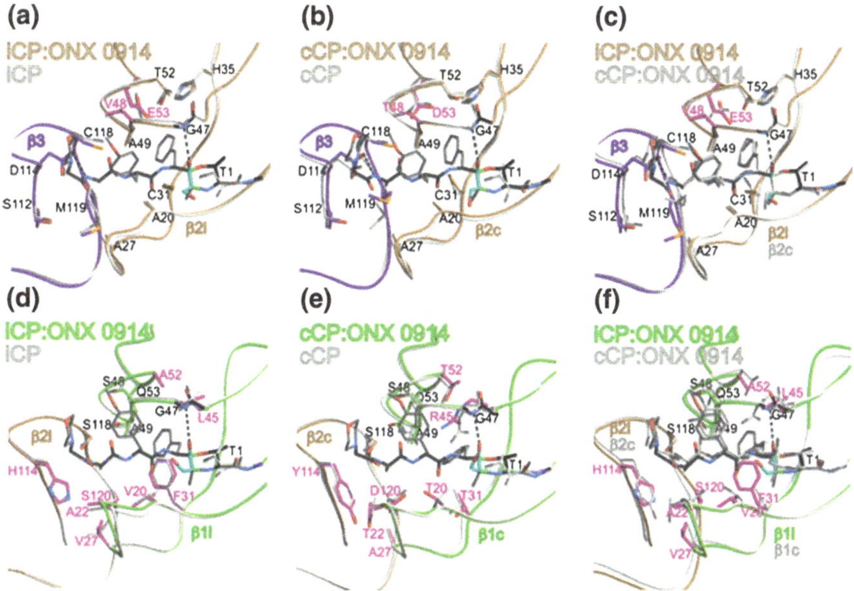

Fig. 4.13 Superposition of ligand-free and ligand-bound active sites of the proteasome. The unliganded subunits β2i (**a**), β2c (**b**), β1i (**d**) and β1c (**e**) are superimposed to their ONX 0914 bound states. The panels (**c**) and (**f**) illustrate the differences for the ligand-bound β2c/i and β1c/i subunits, respectively; amino acid numbers are given for the i subunit. Residues characteristic of the i or c subunits are highlighted in *magenta*. Hydrogen bonds are indicated by *black dashed lines*. Adapted from Huber et al. 2012 [5]

By contrast, binding of ONX 0914 triggers structural rearrangements in subunit β1c. Due to steric hindrance with the ligand's phenylgroup in P1, Arg45 has to change its conformation (Fig. 4.13e). Moreover, the polarity of the S1 pocket disfavours the binding of the hydrophobic P1 side chain of ONX 0914 to subunit β1c and causes electrostatic repulsion.

Therefore, the hydrophobic substrate binding channel of the β1i active site should be ideally suited for docking of ONX 0914. Yet, the ligand complex

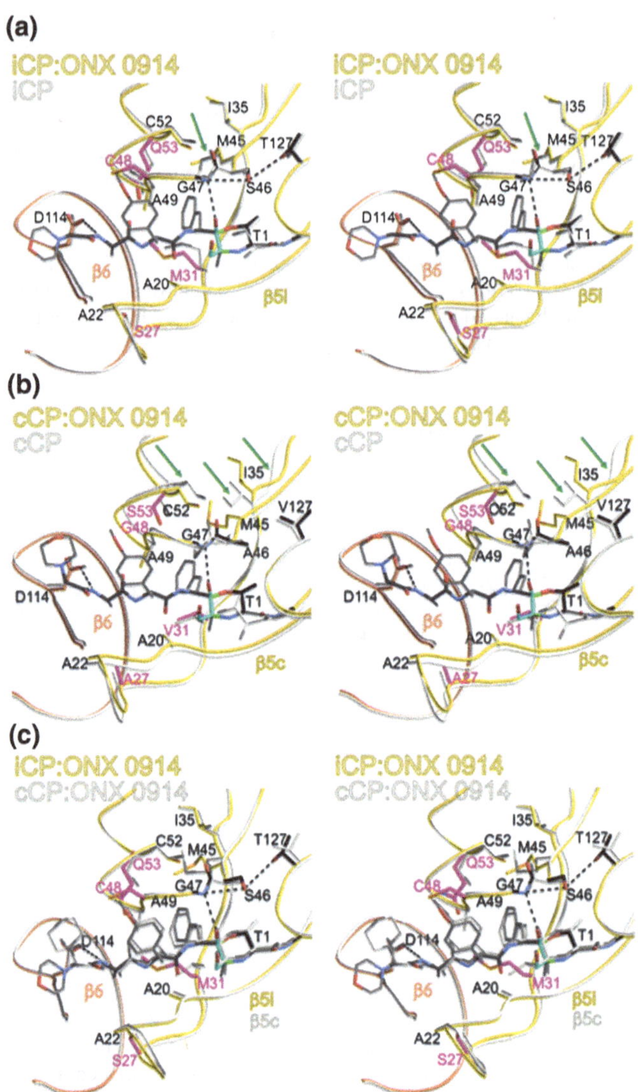

Fig. 4.14 Binding of ONX 0914 to the β5c and β5i subunits of the cCP and iCP (*Stereo view*). (**a**) The unprimed substrate binding channel of subunit β5i is ideally suited to accommodate ligands with bulky P1 side chains such as ONX 0914. Upon binding neighbouring protein residues only slightly rearrange (*green arrow*). (**b**) Docking of ONX 0914 to subunit β5c triggers major structural changes around the active site as marked by *green arrows*. (**c**) The β5c and β5i subunits in their ligand bound states are structurally similar to each other. Adapted from Huber et al. 2012 [5]

structure indicates that Phe31 sterically hinders binding of ONX 0914 to Thr1 (Fig. 4.13d, f). The distance between the phenyl group of Phe31 and the P1 phenylalanine of ONX 0914 is only 3.4–3.5 Å and Phe31 adopts an energetically

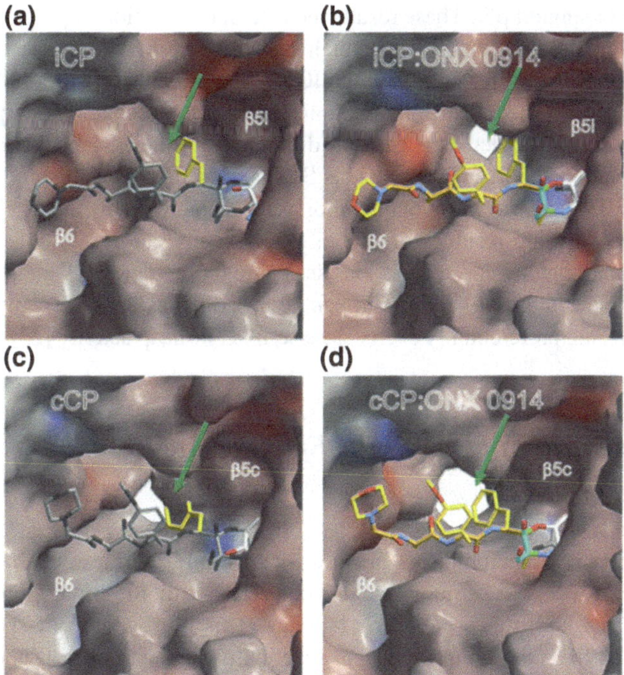

Fig. 4.15 Conolly surface representations of the unprimed β5c and β5i substrate binding channels in the presence and absence of ONX 0914. Surface charge distributions are shown for the subunits β5i (**a**), β5i:ONX 0914 (**b**), β5c (**c**) and β5c:ONX 0914 (**d**) with positive and negative electrostatic potentials contoured from 50 kT/e (*intense blue*) to -50 kT/e (*intense red*). In the panels (**a**) and (**c**) ONX 0914 has been modelled into the active site by superposition with the panels (**b**) and (**d**), respectively. The catalytic Thr1 is coloured in *white*. ONX 0914 is shown in *yellow*, modelled molecules of ONX 0914 in *grey* – except for the P1 site. To depict ligand–protein clashes and resulting structural changes (*green arrows*) the amino acids 46-50 of the subunits β5c and β5i were removed. Adapted from Huber et al. 2012 [5]

disfavoured orientation towards the carbonyl oxygen of Asp32 (distance 3.2 Å). Thus, the atomic distances reflect the repelling forces that prevent high-affinity binding of ONX 0914 to subunit β1i.

Ligand binding to the β5i subunit triggers slight conformational adaptions of Met31 and the CH_3-S-group of Met45 (r.m.s.d. C_α β5i/β5i:ONX 0914: 0.28 Å; Fig. 4.14a). By contrast, comparison of the unprimed substrate binding channel of subunit β5c in the presence and absence of ONX 0914 reveals profound differences. Binding of the bulky P1 phenylalanine of the α’,β’epoxyketone to the S1 pocket displaces Met45 from its normal position. Furthermore, the side chain of Ile35 is flipped and the β sheets S4 and S5 as well as α helix H1 are offset by up to 1.7 Å (r.m.s.d. C_α β5c/β5c:ONX 0914: 0.64 Å; Fig. 4.14b). Remarkably, the ligand bound states of the subunits β5c and β5i are similar to each other (r.m.s.d C_α β5c: ONX 0914/β5i:ONX 0914: 0.55 Å; Fig. 4.14c) and the β5c:ONX 0914 structure highly resembles the apostructure of subunit β5i. These results indicate that inhibition of β5c by ONX 0914 is sterically impaired by Met45. The enthalpic energy required to achieve the observed structural changes is mirrored in the high IC_{50} value of ONX 0914 towards β5c compared to β5i and its lower affinity. Conolly surface representations of the subunits β5c and β5i clearly depict the structural differences in their S1 pocket architecture. Whereas subunit β5i possesses a spacious S1 site, which enables the binding of bulky P1 residues (Fig. 4.15a) with only minor side chain rearrangements (Fig. 4.15b), the S1 pocket forming residues of subunit β5c severely clash with bulky P1 groups (Fig. 4.15c) and thereby cause enormous structural reorientations (Fig. 4.15d). Hence, the substrate specificity of subunit β5i can be described as a ChTL activity, whereas the β5c active site preferentially exerts elastase-like or Snaap activity [23].

4.3.4 Structural Analysis of the Epoxyketone Reaction Mechanism

Whereas boronic acid inhibitors are known to inhibit serine and threonine proteases [24], α',β'epoxyketones exclusively react with the small family of Ntn hydrolases to which the proteasome belongs. This high degree of specificity for the CP results from the unique bivalent reaction mechanism of α',β'epoxyketones, which has been elucidated for the natural product epoxomicin isolated from actinomycetes [25]: The nucleophilic Thr1O^{γ} of the active proteasome subunit attacks the carbonyl carbon atom of the α',β'epoxyketone, thereby forming a reversible hemiketal. Subsequently, the free N-terminus of Thr1 opens the epoxide ring and, by an irreversible cyclisation, creates a morpholine ring system (Fig. 4.16). This mode of action and the docking mechanism of ONX 0914 was investigated in detail by ONX 0914:yCP complex structures. The yCP can be used as a model system due to its structural similarity to the cCP, including the orientation of Met45 (Fig. 4.19a). yCP crystals were therefore soaked with ONX 0914

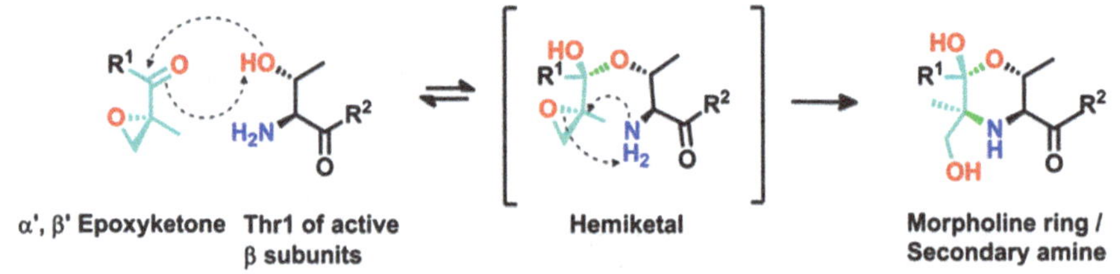

α', β' Epoxyketone Thr1 of active β subunits Hemiketal Morpholine ring / Secondary amine

Fig. 4.16 Schematic illustration of the reaction mechanism of epoxyketone inhibitors. Thr1O^{γ} nucleophilically attacks the carbonyl carbon atom of the epoxyketone inhibitor. The formed hemiketal intermediate can either dissociate to deliberate Thr1O^{γ} or irreversibly cycle under the formation of a secondary amine (morpholine ring system) involving Thr1 N. Bonds created by this reaction are coloured in *green*, R^1 corresponds to the peptide moiety of the compound, R^2 to the proteolytically active β subunit, whose Thr1 is covalently modified by the displayed reaction mechanism. Adapted from Huber and Groll [15]

Table 4.2 X-ray data collection and refinement statistics of yCP structures in complex with ONX 0914

	yCP: ONX 0914_ep	yCP: ONX 0914_mo
Crystal parameters		
Space group	$P2_1$	$P2_1$
Cell constants	a = 135.4 Å	a = 134.4 Å
	b = 300.4 Å	b = 300.8 Å
	c = 143.9 Å	c = 143.8 Å
	β = 112.8°	β = 112.8°
CPs/AU[a]	1	1
Data collection		
Beam line	X06SA, SLS	X06SA, SLS
Wavelength (Å)	1.0	1.0
Resolution range (Å)[b]	30-2.7	30-3.4
	(2.8-2.7)	(3.5-3.4)
No. observations	1211656	443424
No. unique reflections[c]	286910	141633
Completeness (%)[b]	99.1 (98.7)	98.1 (98.3)
R_{merge} (%)[b, d]	9.8 (59.6)	14.3 (59.6)
$^{I}/_{\sigma}\ (I)$[b]	11.3 (2.7)	8.2 (2.2)
Refinement (REFMAC5)		
Resolution range (Å)	15-2.7	15-3.4
No. refl. working set	272564	134550
No. refl. test set	13628	6727
No. non hydrogen	50924	51122
No. of ligand atoms	36	294
Water molecules	1340	1322
R_{work}/R_{free} (%)[e]	22.3/24.3	17.9/22.0
r.m.s.d. bond (Å)/(°)[f]	0.005/0.834	0.005/0.899
Average B-factor ($Å^2$)	54.6	80.5
Ramachandran Plot (%)[g]	97.3/2.4/0.3	96.3/3.2/0.5

[a] Asymmetric unit

[b] The values in parentheses of resolution range, completeness, R_{merge} and $^{I}/_{\sigma}\ (I)$ correspond to the last resolution shell

[c] Friedel pairs were treated as identical reflections

[d] $R_{merge}(I) = \Sigma_{hkl} \Sigma_j \mid [I(hkl)_j - I(hkl)] \mid \Sigma_{hkl} I_{hkl}$, where $I(hkl)_j$ is the jth measurement of the intensity of reflection hkl and $< I(hkl) >$ is the average intensity

[e] $R = \Sigma_{hkl} \mid |F_{obs}| - |F_{calc}| \mid / \Sigma_{hkl} |F_{obs}|$, where R_{free} is calculated for a randomly chosen 5 % of reflections, which were not used for structure refinement, and R_{work} is calculated for the remaining reflections

[f] Deviations from ideal bond lengths/angles

[g] Number of residues in favoured, allowed or outlier region

in different concentrations for varying time periods. Hereby, the reaction intermediate—the hemiketal with an intact epoxide (ep)—and the reaction product—the morpholine ring (mo)—could be trapped in a crystal structure (Table 4.2). These data clarified that indeed Thr1 N attacks the epoxide once the hemiketal is formed, and proved the formerly proposed mode of action of epoxyketone inhibitors [25].

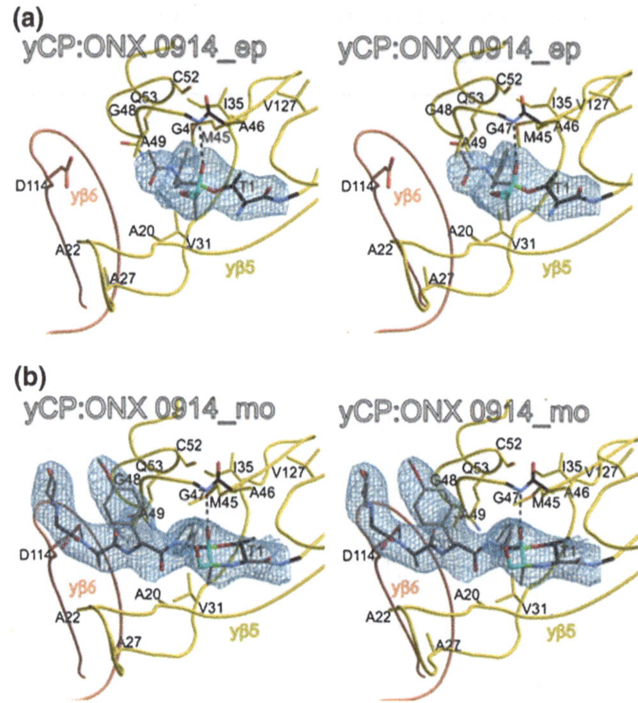

Fig. 4.17 *Stereo view* of the yeast 20S proteasome subunit β5 (yβ5) in complex with ONX 0914. Subunit yβ5 is coloured in *yellow* and the neighbouring subunit yβ6 in *orange*. Thr1 and the oxyanion hole Gly47NH are marked in *black*; and hydrogen bonds are indicated by *black dashed lines*. ONX 0914 is shown in *grey* with its reactive head group highlighted in *cyan*. The experimental $2F_O$-F_C omit electron density map (*blue* mask) is contoured to 1σ. (**a**) Hemiketal formation of ONX 0914 with Thr1 of subunit yβ5. Note, only the epoxide and the P1 site of ONX 914 are well-defined in the electron density map. (**b**) Subunit yβ5 in complex with the completely reacted ONX 0914. All inhibitor side chains (P1, P2 and P3) are visualized in the electron density. Adapted from Huber et al. 2012 [5]

Whereas all three proteolytically active sites were covalently modified in the yCP:ONX 0914_mo structure, only subunit yβ5 visualized the hemiketal formation of ONX 0914 (Fig. 4.17a). In agreement with data on the murine and human cCP [19], ONX 0914 favours subunit yβ5 over yβ1 and yβ2.

Notably, the $2F_O$-F_C electron density map for the yCP:ONX 0914_mo structure displayed the whole inhibitor with its P1, P2 and P3 residues as well as the N-terminal morpholine ring and the covalent linkage to Thr1 (Fig. 4.17b). In contrast, the electron density map for the hemiketal intermediate (yCP:ONX 0914_ep) surprisingly depicted solely the intact epoxide ring and the ligand's P1 site (Fig. 4.17a). As the N-terminal P3 and P2 sites of ONX 0914_ep were not defined, they were not yet bound to their respective substrate specificity pockets and thus, flexible. This finding provides evidence that first the electrophilic warhead and the P1 residue of covalently acting inhibitors contact the active site Thr1 and the S1

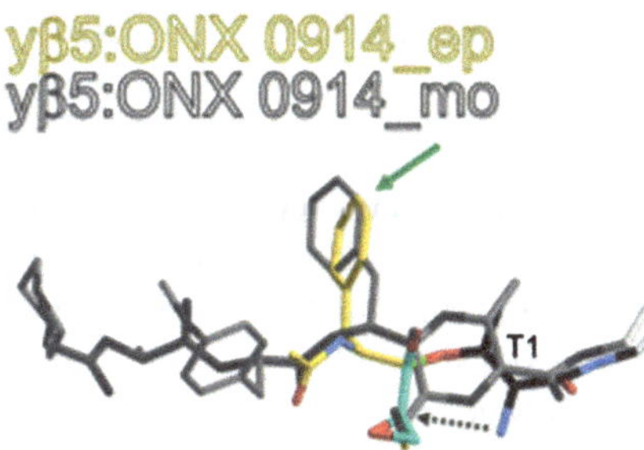

Fig. 4.18 Docking of ONX 0914 to the active site Thr1. Structural superposition of ONX 0914 in its epoxide (ep; *yellow*) and morpholine (mo; *grey*) state bound to Thr1 of subunit yβ5. The green arrow indicates that the positions of the P1 side chain differ in the reaction intermediate and the reaction product; the black dashed arrow marks the position where the free N-terminus of yβ5 attacks the epoxide to form a morpholine ring system. Adapted from Huber et al. 2012 [5]

pocket and that only upon completing the reaction mechanism interactions with the S2 and S3 sites form the antiparallel β sheet in the substrate binding channel. In conclusion, even though all substrate binding pockets contribute to the IC_{50} value of a compound, the P1 side chain of a compound, apart from its electrophilic pharmacophore, is mostly responsible for its affinity for the active site. In this regard the depicted size differences of the β5c and β5i S1 pockets provoke the enhanced affinity of ONX 0914 for subunit β5i.

Structural comparison of the hemiketal intermediate and the final reaction product reveals that although the P1 residue of the ONX 0914_ep is defined in the electron density map, it initially does not fully occupy the S1 pocket of subunit yβ5 (Fig. 4.18). Hence, during the reversible hemiketal formation no sterical clashes with Met45 occur and in agreement with the X-ray structure of the intermediate state no displacement of Met45 is observed (Fig. 4.17a). Only the formation of the morpholine ring system with Thr1 properly positions the P1 side chain of ONX 0914 in the S1 specificity site and triggers the reorientation of Met45 as well as further structural changes (Fig. 4.17b). With respect to the distinct affinities of ONX 0914 for the subunits β5c and β5i, the following model for ONX 0914 binding to cCP and iCP β5 subunits is proposed: In both active sites β5i and β5c hemiketal formation of ONX 0914 with Thr1O^{γ} is possible without any sterical hindrance. However, whereas the P1 side chain of ONX 0914 can immediately adopt its final position in the S1 pocket of subunit β5i and thereby form the morpholine ring system as well as the antiparallel β sheet, this is not possible in subunit β5c. Met45 of subunit β5c sterically hampers full binding of the P1 residue to the S1 pocket and has to be dislocated to enable covalent modification of Thr1. Consequently, the energy barrier for the morpholine ring formation in subunit β5c is higher and the back-reaction leading to a restored and catalytically active Thr1 in subunit β5c is more likely than in β5i. Thus, the increased probability of ONX 0914 to modify the Thr1 of subunit β5i compared to β5c provides the explanation for its lower IC_{50} value and its selectivity for subunit β5i.

4.4 Yeast Mutagenesis Studies on the Subunits β5c, β5i and β5t

The crystal structures of the murine cCP and iCP demonstrated that the proteasomal β subunits are structurally conserved from archaea to mammals. Based on the overall identical folds and similar main chain tracings of β5 subunits (Fig. 4.19), the yCP can be used as a model system for analysing the functional impact of single amino acid exchanges between the cCP, iCP as well as tCP by mutagenesis.

4.4.1 Mimicking the β5c Active Site

The cCP and iCP crystal structures depicted that the orientation of Met45 is crucial for the size of the S1 specificity pocket in β5 subunits (Figs. 4.9, 4.14). While favourable van der Waals interactions of Met45 with Gln53 contribute to the formation of a large S1 site in subunit β5i, Ser53 cannot stabilize Met45 in subunit β5c (Fig. 4.9) and causes a diminished S1 pocket in subunit β5c. This model is supported by the strong conservation of Ser53 in β5c subunits, Gln53 in β5i entities and Ala53 in β5t active sites [5]. Notably, the yeast yβ5 subunit incorporates Lys32, a hallmark of β5c entities, as well as the β5i characteristic residue Gln53, leading to a similar S1 pocket architecture like in β5c. Hence, subunit yβ5 appears to represent a chimera of the β5c and β5i active sites at least with respect to the amino acids 32 and 53. Indeed, the IC_{50} value of ONX 0914 for subunit yβ5 was determined to ~0.5 μM, while it was ~1 μM for β5c and ~0.07 μM for β5i (Fig. 4.10).

The importance of Gln53 for the S1 pocket architecture in the subunits yβ5 and β5i was proven by mutation of Gln53 to Ser in the yβ5 active site. The IC_{50} values of ONX 0914 and bortezomib for the created mutant CP were 1.32 μM and 0.15 μM, respectively (Fig. 4.20). Remarkably, compared to the wt yCP, ONX 0914 was three-times less potent towards the yβ5 Q53S mutant, while bortezomib displayed an increased affinity (Fig. 4.20). These findings elucidate that Ser53 enhances binding of ligands with smaller P1 side chains such as Leu (bortezomib) over inhibitors with bulky residues like Phe (ONX 0914). Wt yβ5 shows only slight preference for bortezomib, but the mutation Q53S significantly increases the selectivity for this compound (Fig. 4.20). In agreement, subunit β5c was reported to be far more susceptible to inhibition by bortezomib (IC_{50}: 7 nM [26]) than ONX 0914 (IC_{50}: 236-1000 nM depending on the method [5, 19]). For structural analysis of the Q53S mutant X-ray data were collected in the presence and the absence of both ligands (Table A.2). The mutant yβ5 subunit Q53S displays high structural identity to its wt counterpart (r.m.s.d. C_α yβ5/yβ5Q53S: 0.2 Å; Fig. 4.21a) and binding of bortezomib and ONX 0914 induces structural changes similar to yβ5 and β5c (r.m.s.d. C_α yβ5Q53S/yβ5Q53S:ONX 0914: 0.76 Å; r.m.s.d. C_α yβ5Q53S/yβ5Q53S: bortezomib: 0.62 Å; Fig. 4.21b, c, f). However, ONX 0914 leads to a more severe displacement of the sulphur atom of Met45 than bortezomib (Fig. 4.21d), hereby explaining the reduced affinity of ONX 0914 for subunit yβ5Q53S compared to bortezomib (Fig. 4.20).

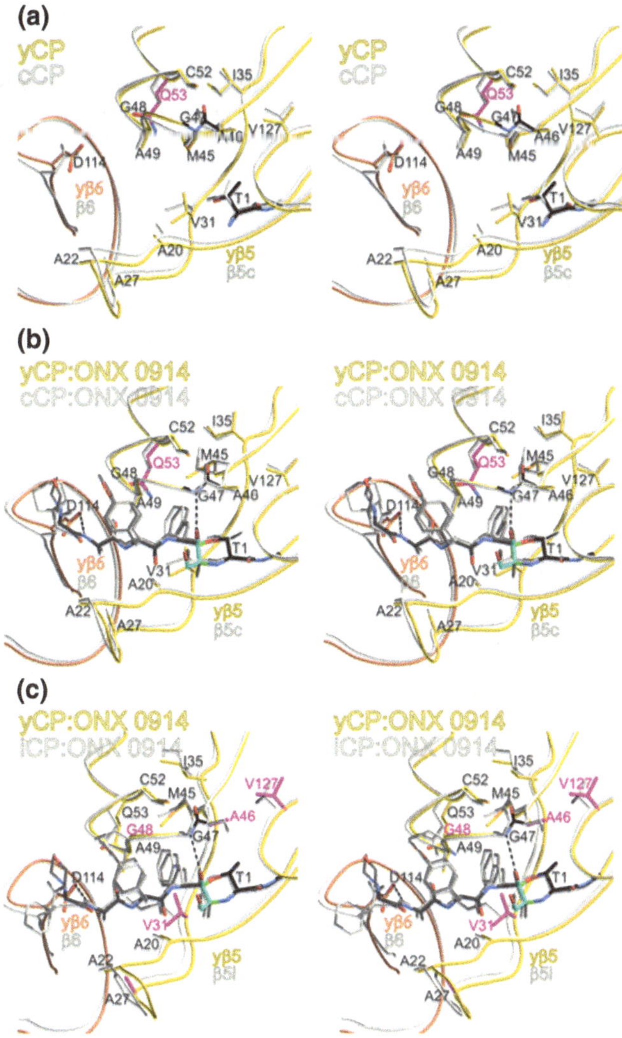

Fig. 4.19 Structural comparison of the yeast subunit yβ5 and the murine active sites β5c and β5i of cCP and iCP, respectively (*stereo view*). (**a**) Superposition of the subunits yβ5 and β5c in their ligand-free states and (**b**) their ligand-bound states depict the structural similarity of both. Panel (**c**) additionally illustrates the comparison of the ONX 0914 bound active sites yβ5 and β5i. Amino acids are numbered for the yCP. Residues that are characteristic of the yCP subunit yβ5 are coloured in *magenta*. Gly47 and Thr1 are shown in *black*. The ligand is highlighted in *grey*. Hydrogen bonds are indicated by *black dashed lines*. Adapted from Huber et al. 2012 [5]

Removal of Gln53 leads to an insufficient stabilization of Met45 in the ligand-bound state and hence, more energy is required for bulky P1 side chains to reorient Met45 and to keep distance to it. The propensity of Met45 to adopt its normal conformation is likely to hinder particularly spacious ligands such as ONX 0914 from binding

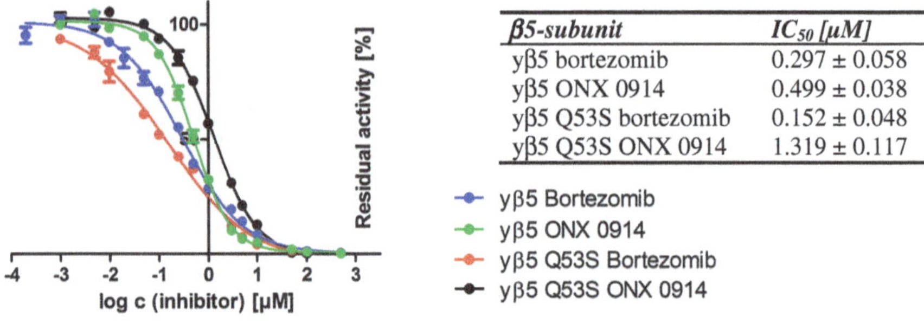

β5-subunit	*IC_{50} [µM]*
yβ5 bortezomib	0.297 ± 0.058
yβ5 ONX 0914	0.499 ± 0.038
yβ5 Q53S bortezomib	0.152 ± 0.048
yβ5 Q53S ONX 0914	1.319 ± 0.117

Fig. 4.20 Inhibition of wt and mutant yCPs by ONX 0914 and bortezomib. Residual proteolytic activities were measured in triplicate upon exposure to different inhibitor concentrations with the fluorogenic substrate Suc-Leu-Leu-Val-Tyr-AMC and normalized to a DMSO treated control. IC_{50} values were deduced from fitted data

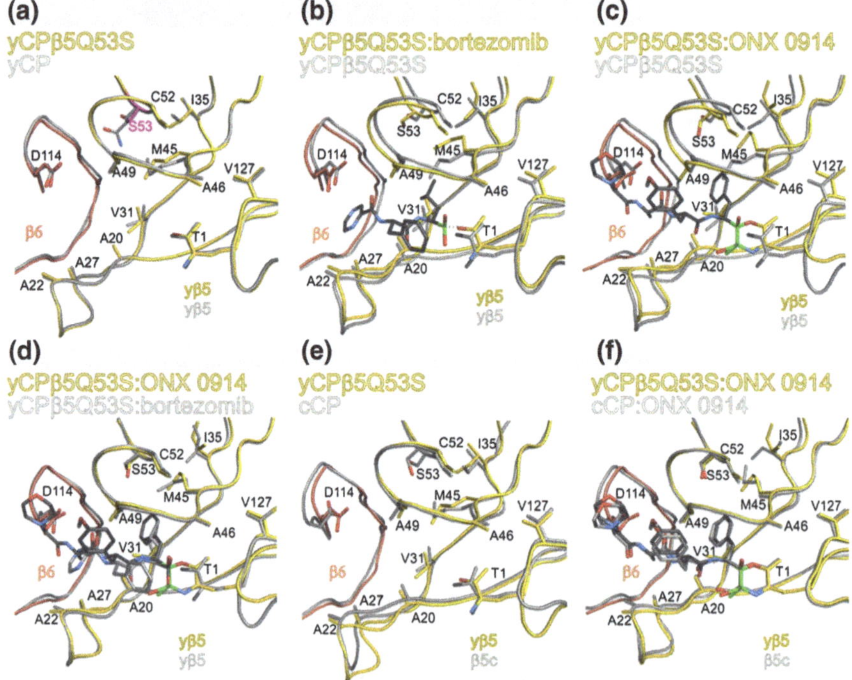

Fig. 4.21 Structural analysis of the yCP β5 mutant Q53S. (**a–d**) Superposition of (un-)liganded wt and mutant yeast β5 active sites; (**e, f**) Comparison of the yCP β5 mutant Q53S with the murine cCP. Amino acid numbers are given for the yCP β5 mutant Q53S; sequence differences between superimposed structures are highlighted in *magenta*; the electrophilic head groups of inhibitory compounds (*grey*) are coloured in *green*

irreversibly to Thr1. In conclusion, although the substitution Q53S causes no obvious structural differences and amino acid 53 is not directly involved in ligand binding, its side chain length yet significantly affects the affinity of ligands.

4.4.2 Analysis of the β5i Substrate Binding Channel

The comparison of the crystal structures of cCP and iCP suggests that van der Waals interactions of Met45 with the aliphatic side chain of Gln53 in subunit β5i promote the formation of a spacious S1 pocket (Fig. 4.9). Although subunit yβ5 harbours a Gln53, Met45 adopts a conformation similar to subunit β5c. This observation provoked the question of which additional amino acids in proximity to the active site contribute to the distinct conformations of Met45.

To identify key residues that lead to the differently sized S1 pocket in the β5i substrate binding channel, numerous point mutations were introduced in subunit yβ5. The following amino acid exchanges aimed at mimicking the murine (or human) β5i subunit: A27S, (V31M,) K32N, A46S, G48C, T57R, K71G and V127T (Fig. 4.1; Table 3.3).

Remarkably, most mutant yeast strains suffer from a significant growth phenotype, which is probably caused by an impaired β5 activity, as observed in a proteolysis assay with the chromogenic substrate Cbz-Gly-Gly-Leu-pNA (Fig. 4.22a). Hereby, the single point mutation G48C sufficed to markedly attenuate the ChTL activity. Cys48 might restrict the flexibility of the loop segment 46-49 and thereby hamper substrate and inhibitor binding. While the exchange A46S alone did not affect CP activity and its combination with G48C had only minor effects on the IC_{50} value of ONX 0914, additional introduction of Thr127 strongly impaired ligand binding (Table 4.3). The hydrogen bond between Ser46 and Thr127 that has also been observed for the iCP structure additionally enhances the rigidity of the loop region 46-49 and thereby might constrict ligand binding. Intriguingly, mutagenesis of additional residues (A27S, K32N, V31M, T57R, K71G) had neither further influence on the affinity of ONX 0914 for the yβ5 active site (Fig. 4.22b) and nor did it lead to the reorientation of Met45 as observed in the unliganded subunit β5i (Fig. 4.23e; Table A.3).

Table 4.3 IC_{50} values for yCP mutants that partially mimic either the primary sequence or the structural features of the iCP subunit β5i. Mutants are termed according to Table 3.3

β5-subunit	Inhibitor	IC_{50} [μM]
yβ5	bortezomib	0.297 ± 0.058
yβ5	ONX 0914	0.499 ± 0.038
yβ5 SC	ONX 0914	0.673 ± 0.055
yβ5 SCT	ONX 0914	5.379 ± 1.839
yβ5 SMSCT	ONX 0914	4.416 ± 0.771
yβ5 SNSCRGT	bortezomib	1.807 ± 0.168
yβ5 SNSCRGT	ONX 0914	5.580 ± 0.307
yβ5 SNASCRGT	bortezomib	3.078 ± 0.402
yβ5 SNASCRGT	ONX 0914	5.706 ± 0.321
yβ5 SNVSCRGT	bortezomib	2.606 ± 0.266
yβ5 SNVSCRGT	ONX 0914	4.911 ± 0.409
yβ5 R	bortezomib	2.508 ± 0.178
yβ5 R	ONX 0914	5.185 ± 1.097
yβ5 TR	bortezomib	0.706 ± 0.115
yβ5 TR	ONX 0914	10.37 ± 2.134

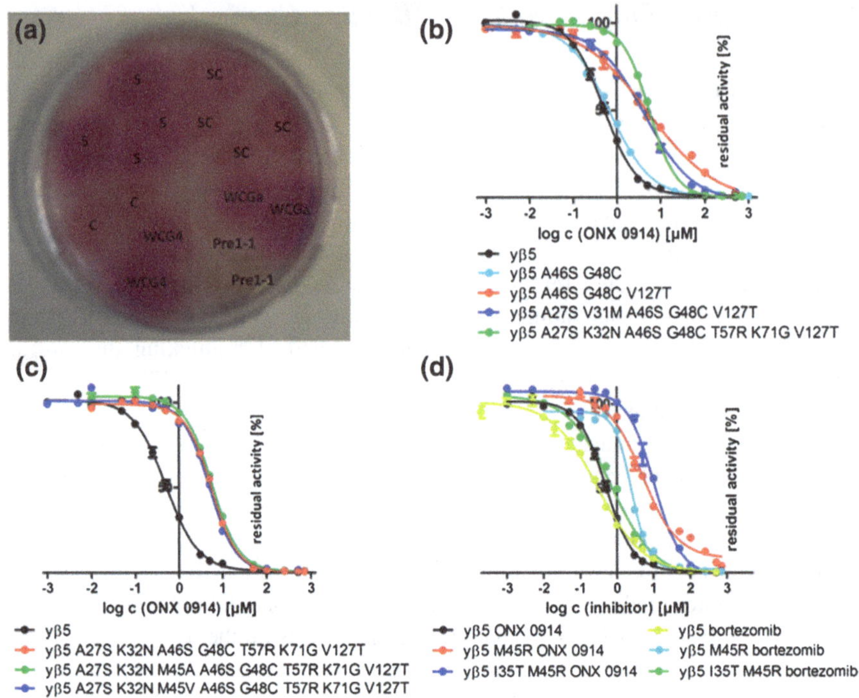

Fig. 4.22 Activity and inhibition assays of mutant and wt yCPs. (**a**) Overlay test for the ChTL activity of yCPs using the chromogenic substrate Cbz-Gly-Gly-Leu-pNA. Tested mutants are termed according to Table 3.3. The yeast strains WCG4a and WCG4a *pre1-1* served as positive and negative controls, respectively. (**b–d**) The indicated mutant CPs that aimed at imitating the iCP subunit β5i were tested in triplicate for their inhibition by ONX 0914 and bortezomib with the fluorogenic substrate Suc-Leu-Leu-Val-Tyr-AMC. The deduced IC_{50} values are given in Table 4.3

Hence, the key features of subunit β5i could so far not be reconstituted in yeast, even though all amino acids in the active site surrounding were identical to the murine/human subunit β5i (Fig. 4.23e). Consequently, rather the structural differences than the primary sequences of the β5i and β5c/yβ5 substrate binding channels are responsible for the selectivity of ONX 0914. In order to mimic the enlarged S1 pocket of subunit β5i, Met45 was substituted by either alanine or valine, but intriguingly these mutant CPs exhibited only poor affinity for both ONX 0914 and bortezomib (Table 4.3). Probably in the absence of a prolonged linear aliphatic side chain such as Met45 ligands cannot be sufficiently stabilized in the S1 pocket. These results are also in line with reports on the bortezomib resistance conferring character of Val45 [27] and are indicative of the highly sophisticated architecture of the proteasome.

Concomitantly, structural characterization of the yβ5 mutants M45R and I35T M45R that were previously analysed [28] and that are unrelated to the iCP revealed an interesting feature. The amino acid exchange M45R reduces the ChTL activity by electrostatic repulsion of apolar amino acids by Arg45, but by the additional

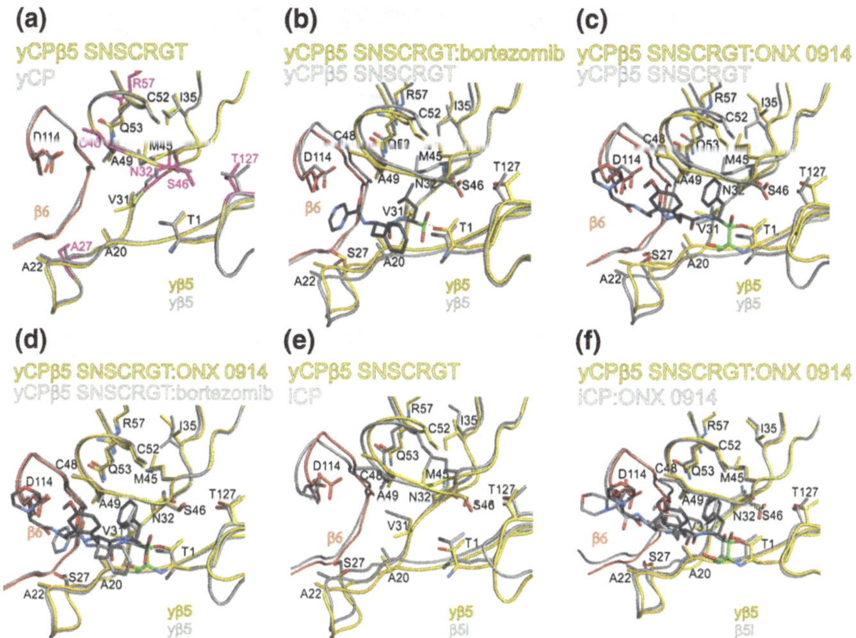

Fig. 4.23 Crystal structure analysis of β5i mimicking yCPs. Mutant yβ5 substrate binding channels are superimposed onto wt β5 active sites of wt yCP and iCP to depict structural similarities and differences. Amino acid labels are indicated for mutant CPs; amino acid mutations are highlighted in *magenta* and the electrophilic headgroups of inhibitors are shown in *green*

mutation I35T the ChTL activity is regained. Ligand complex structures of the mutant M45R with bortezomib and ONX 0914 visualized that only binding of the α',β'epoxyketone causes major structural changes of Arg45 (r.m.s.d. C_α yβ5M45R/yβ5M45R:ONX 0914: 0.82 Å; r.m.s.d. C_α yβ5M45R/yβ5M45R:bortezomib: 0.43 Å; Fig. 4.24b, c; Table A.4). Hence, the poor affinity of ONX 0914 is presumably caused by sterical hindrance and the opposing polarity in the S1 pocket, while binding of bortezomib appears to be solely hampered by electrostatic repulsion.

Interestingly, the restored ChTL activity of the double mutant relies on a strong hydrogen bond (2.8 Å) between Thr35O^γ and Arg45N^ε that rearranges Arg45 and enlarges the size of the S1 pocket similar to β5i (Fig. 4.24g, k). Due to the spacious S1 site the double mutant displayed a 3.5 times enhanced affinity for bortezomib compared to the yβ5M45R mutant (Table 4.3) and the ligand does not induce any structural rearrangement (r.m.s.d.C_αyβ5TR/yβ5TR:bortezomib: 0.35 Å; Fig. 4.24h; Table A.5). This finding proves that indeed the size of the S1 pocket affects the inhibitory potency of compounds. Still, the mutant yCP I35T M45R is dramatically less susceptible to ONX 0914 than the wt yCP (Table 4.3), because accommodation of its bulky phenyl side chain is electrostatically hindered by the guanidine group of Arg45 (Fig. 4.24i). As confirmed by crystallographic analysis ligand binding pushes Arg45 1 Å further away from the phenylgroup of

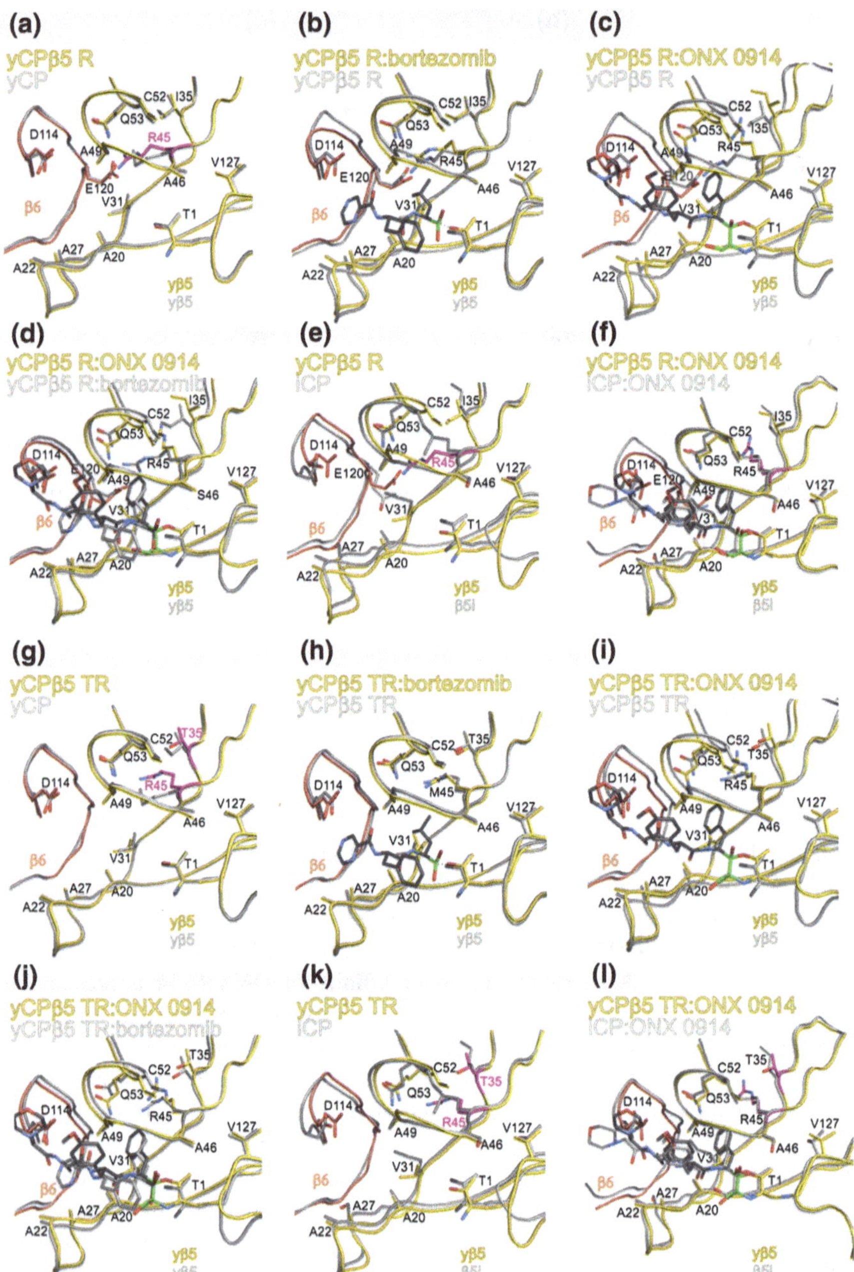

Fig. 4.24 Crystal structures of mutant CPs. Mutant yβ5 substrate binding channels are superimposed onto β5 active sites of wt yCP and iCP to depict structural similarities and differences in the presence and absence of inhibitor. Amino acid residues of mutant CPs are labelled; amino acid mutations are highlighted in *magenta* and the electrophilic headgroups of inhibitors are shown in *green*

ONX 0914, thereby significantly enlarging its distance to Arg45 up to 5 Å (r.m.s.d. C_α yβ5TR/yβ5TR:ONX 0914: 0.69 Å; Fig. 4.24i). A similar distance of 4.8-5.2 Å is observed for Arg45 and the P1 leucine of bortezomib bound to the yCP I35T M45R mutant (Fig. 4.24h). In conclusion, the strongly reduced affinity of ONX 0914 for the double mutant discloses the enormous effects of electrostatic repulsion on ligand binding (see also Sect. 4.4.3) and highlights the perfect suitability of the Met side chain for establishing a ChTL activity.

4.4.3 Probing Subunit β5t

The thymoproteasome incorporates the tCP-specific subunit β5t that—unlike the β5c and β5i active sites—harbours a more hydrophilic substrate binding channel [6]. In particular the amino acid residues Ser20, Ser31, Thr45, Thr48 and Ser49 (Fig. 4.1) enhance the polarity of the unprimed S1 and S2 pockets and favour cleavage C-terminally of charged residues. In contrast, the S3 pocket is lined with the hydrophobic residues Cys22 and Ala27. Similarly to subunit β5i, Ser46 might increase the polarity around the active site Thr1 and the oxyanion hole (see also Sect. 4.2.3). Val127, however, is not suited to form a hydrogen bond network with Ser46 like it is observed for subunit β5i.

Due to limited biological samples no structural information on the tCP is available so far. Therefore, this study aimed at mimicking the β5t active site in yeast by mutagenesis. A stepwise approach was used to introduce the mutations A20S, A22C, V31S, M45T, A46S and G48T into subunit yβ5 (Table 3.3). Most of the created mutants were characterized by a reduced growth rate and displayed a dramatically decreased ChTL activity of the yβ5 subunit (Fig. 4.25a). Moreover, the IC_{50} value of ONX 0914 for the SCSTST mutant yβ5 active site strongly increased to 20.57 ± 3.78 μM compared to 0.499 μM for wt yβ5. Even though the S1 pocket of the yβ5 SCSTST mutant is enlarged, the opposing forces between the hydrophobic side chains of ONX 0914 and the charged amino acid lining of the substrate binding pockets of the yβ5 mutant SCSTST strongly impair binding of the ligand.

Despite the reduced affinity of ONX 0914 for this mutant yCP structural data could be obtained in the absence and the presence of the compound (Table A.6). Positive F_O-F_C density maps proved the incorporation of all mutations. The wt and mutant yβ5 active sites superimpose well and show no structural changes for the mutant subunit (r.m.s.d. C_α yβ5/yβ5t: 0.2 Å). Comparison of the ligand-free and ligand-bound states of the β5t-like substrate binding channel depicts the turn away of Thr45 from the hydrophobic ligand ONX 0914. In addition, strong backbone distortions (r.m.s.d. C_α yβ5t/yβ5t:ONX 0914: 0.72 Å) are observed upon ligand binding even though no obvious clashes with the protein occur. The reason for this tremendous backbone shift upon inhibitor binding cannot be deduced from the mutant structures and has to be further investigated by structural analysis of the tCP.

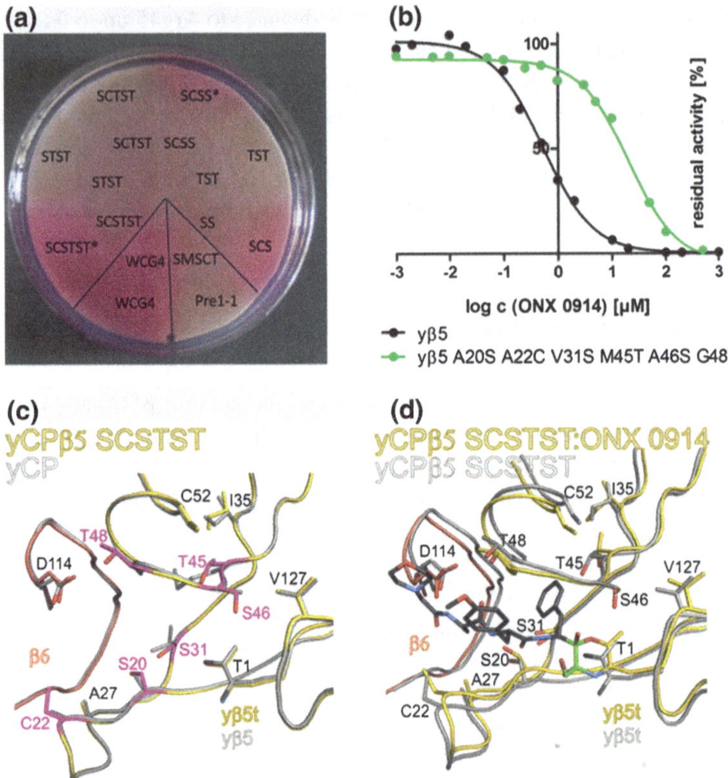

Fig. 4.25 Structural examination of a β5t mimicking yCP. (**a**) An overlay assay using the chromogenic substrate Cbz-Gly-Gly-Leu-pNA shows the marked reduction in ChTL activity upon introduction of β5t-specific mutations. Wildtype yeast (WCG4a) served as a positive control, whereas the β5i-mimicking yβ5 mutant SMSCT and the CP mutant Pre1-1, which is lacking ChTL activity, were used as negative controls. Mutant clones marked with a star showed unspecific staining. For abbreviations of mutants see Table 3.3 (**b**) Suc-Leu-Leu-Val-Tyr-AMC was used as a fluorogenic substrate to assess the inhibition of the mutated subunit β5 by the epoxyketone ONX 0914. The affinity of ONX 0914 is strongly attenuated in the mutant CP (**c, d**) Comparison of the wildtype and mutant (SCSTST) yβ5 substrate binding channels depicts the differences in their primary sequences (magenta) and illustrates structural changes upon ligand binding. Amino acids are labelled for the mutant CP

References

1. J. Löwe, D. Stock, B. Jap, P. Zwickl, W. Baumeister, R. Huber, Crystal structure of the 20S proteasome from the archaeon *T. acidophilum* at 3.4 Å resolution. Science **268**, 533–539 (1995)
2. M. Groll, L. Ditzel, J. Löwe, D. Stock, M. Bochtler, H.D. Bartunik, R. Huber, Structure of 20S proteasome from yeast at 2.4 Å resolution. Nature **386**, 463–471 (1997)
3. P. Zwickl, A. Grziwa, G. Puhler, B. Dahlmann, F. Lottspeich, W. Baumeister, Primary structure of the Thermoplasma proteasome and its implications for the structure, function, and evolution of the multicatalytic proteinase. Biochemistry **31**, 964–972 (1992)
4. M. Unno, T. Mizushima, Y. Morimoto, Y. Tomisugi, K. Tanaka, N. Yasuoka, T. Tsukihara, The structure of the mammalian 20S proteasome at 2.75 Å resolution. Structure **10**, 609–618 (2002)

5. E. Huber, M. Basler, R. Schwab, W. Heinemeyer, C.J. Kirk, M. Groettrup, M. Groll, Immuno- and constitutive proteasome crystal structures reveal differences in substrate and inhibitor specificity. Cell **148**, 727–738 (2012)
6. S. Murata, K. Sasaki, T. Kishimoto, S. Niwa, H. Hayashi, Y. Takahama, K. Tanaka, Regulation of CD8 + T cell development by thymus-specific proteasomes. Science **316**, 1349–1353 (2007)
7. S. Murata, Y. Takahama, K. Tanaka, Thymoproteasome: probable role in generating positively selecting peptides. Curr. Opin. Immunol. **20**, 192–196 (2008)
8. G. Schmidtke, S. Emch, M. Groettrup, H.G. Holzhutter, Evidence for the existence of a non-catalytic modifier site of peptide hydrolysis by the 20 S proteasome. J. Biol. Chem. **275**, 22056–22063 (2000)
9. M. Groettrup, R. Kraft, S. Kostka, S. Standera, R. Stohwasser, P.M. Kloetzel, A third interferon-gamma-induced subunit exchange in the 20S proteasome. Eur. J. Immunol. **26**, 863–869 (1996)
10. M. Groll, M. Bajorek, A. Köhler, L. Moroder, D.M. Rubin, R. Huber, M.H. Glickman, D. Finley, A gated channel into the proteasome core particle. Nat. Struct. Biol. **7**, 1062–1067 (2000)
11. F.G. Whitby, E.I. Masters, L. Kramer, J.R. Knowlton, Y. Yao, C.C. Wang, C.P. Hill, Structural basis for the activation of 20S proteasomes by 11S regulators. Nature **408**, 115–120 (2000)
12. P.C. Ramos, A.J. Marques, M.K. London, R.J. Dohmen, Role of C-terminal extensions of subunits beta2 and beta7 in assembly and activity of eukaryotic proteasomes. J. Biol. Chem. **279**, 14323–14330 (2004)
13. B. Guillaume, J. Chapiro, V. Stroobant, D. Colau, B. Van Holle, G. Parvizi, M.P. Bousquet-Dubouch, I. Theate, N. Parmentier, B.J. Van den Eynde, Two abundant proteasome subtypes that uniquely process some antigens presented by HLA class I molecules. Proc. Natl. Acad. Sci. U S A **107**, 18599–18604 (2010)
14. N. Klare, M. Seeger, K. Janek, P.R. Jungblut, B. Dahlmann, Intermediate-type 20 S proteasomes in HeLa cells: "asymmetric" subunit composition, diversity and adaptation. J. Mol. Biol. **373**, 1–10 (2007)
15. E.M. Huber, M. Groll, Inhibitors for the immuno- and constitutive proteasome: current and future trends in drug development. Angew. Chem. Int. Ed. Engl. **51**, 8708–8720 (2012)
16. M.A. Gräwert, M. Groll, Exploiting nature's rich source of proteasome inhibitors as starting points in drug development. Chem. Comm. **48**, 1364–1378 (2012)
17. P. Beck, C. Dubiella, M. Groll, Covalent and non-covalent reversible proteasome inhibition. Biol. Chem. **393**, 1101–1120 (2012)
18. C. Blackburn, K.M. Gigstad, P. Hales, K. Garcia, M. Jones, F.J. Bruzzese, C. Barrett, J.X. Liu, T.A. Soucy, D.S. Sappal, N. Bump, E.J. Olhava, P. Fleming, L.R. Dick, C. Tsu, M.D. Sintchak, J.L. Blank, Characterization of a new series of non-covalent proteasome inhibitors with exquisite potency and selectivity for the 20S beta5-subunit. Biochem. J. **430**, 461–476 (2010)
19. T. Muchamuel, M. Basler, M.A. Aujay, E. Suzuki, K.W. Kalim, C. Lauer, C. Sylvain, E.R. Ring, J. Shields, J. Jiang, P. Shwonek, F. Parlati, S.D. Demo, M.K. Bennett, C.J. Kirk, M. Groettrup, A selective inhibitor of the immunoproteasome subunit LMP7 blocks cytokine production and attenuates progression of experimental arthritis. Nat. Med. **15**, 781–787 (2009)
20. H.T. Ichikawa, T. Conley, T. Muchamuel, J. Jiang, S. Lee, T. Owen, J. Barnard, S. Nevarez, B.I. Goldman, C.J. Kirk, R.J. Looney, J.H. Anolik, Novel proteasome inhibitors have a beneficial effect in murine lupus via the dual inhibition of type i interferon and autoantibody secreting cells. Arthritis Rheum. **64**, 493–503 (2011)
21. M. Basler, M. Dajee, C. Moll, M. Groettrup, C.J. Kirk, Prevention of experimental colitis by a selective inhibitor of the immunoproteasome. J. Immunol. **185**, 634–641 (2010)
22. Y. Nagayama, M. Nakahara, M. Shimamura, I. Horie, K. Arima, N. Abiru, Prophylactic and therapeutic efficacies of a selective inhibitor of the immunoproteasome for Hashimoto's thyroiditis, but not for Graves' hyperthyroidism, in mice. Clin. Exp. Immunol. **168**, 268–273 (2012)
23. M. Orlowski, C. Cardozo, C. Michaud, Evidence for the presence of five distinct proteolytic components in the pituitary multicatalytic proteinase complex. Properties of two components cleaving bonds on the carboxyl side of branched chain and small neutral amino acids. Biochemistry **32**, 1563–1572 (1993)
24. S. Arastu-Kapur, J.L. Anderl, M. Kraus, F. Parlati, K.D. Shenk, S.J. Lee, T. Muchamuel, M.K. Bennett, C. Driessen, A.J. Ball, C.J. Kirk, Nonproteasomal targets of the proteasome

inhibitors bortezomib and carfilzomib: a link to clinical adverse events. Clin. Cancer Res. **17**, 2734–2743 (2011)
25. M. Groll, K.B. Kim, N. Kairies, R. Huber, C.M. Crews, Crystal structure of epoxomicin: 20S proteasome reveals a molecular basis for selectivity of α', β'-epoxyketone proteasome inhibitors. J. Am. Chem. Soc. **122**, 1237–1238 (2000)
26. S.D. Demo, C.J. Kirk, M.A. Aujay, T.J. Buchholz, M. Dajee, M.N. Ho, J. Jiang, G.J. Laidig, E.R. Lewis, F. Parlati, K.D. Shenk, M.S. Smyth, C.M. Sun, M.K. Vallone, T.M. Woo, C.J. Molineaux, M.K. Bennett, Antitumor activity of PR-171, a novel irreversible inhibitor of the proteasome. Cancer Res. **67**, 6383–6391 (2007)
27. N.E. Franke, D. Niewerth, Y.G. Assaraf, J. van Meerloo, K. Vojtekova, C.H. van Zantwijk, S. Zweegman, E.T. Chan, C.J. Kirk, D.P. Geerke, A.D. Schimmer, G.J. Kaspers, G. Jansen, J. Cloos, Impaired bortezomib binding to mutant beta5 subunit of the proteasome is the underlying basis for bortezomib resistance in leukemia cells. Leukemia **26**, 757–768 (2011)
28. R.J.C. Estiveira, The active subunits of the 20S Proteasome in *Saccharomyces cerevisiae*—Mutational analysis of their specificities and a C-terminal extension, PhD thesis, Universität Stuttgart, 2008

Chapter 5
Discussion

The here presented crystallographic analysis of the murine cCP and iCP emphasizes the biological impact of subtle differences that are not predictable from sequence alignments and that can only be resolved by structural data. However, owing to the applied methodology, the results presented provide no insights into the dynamics of the 20S proteasome that might have substantial effects on substrate binding as well as enzyme inhibition.

5.1 Structural and Functional Differences Between the Three Types of CPs

The high structural identity of the unprimed β2c and β2i substrate binding channels implicates similar cleavage preferences, leading to the generation of MHC I ligands with neutral or basic C-terminal amino acids [1]. Hence, the cCP and iCP structures elicit the question of why β2i, which is the only i subunit that is not encoded on the MHC gene cluster, is sequestered into iCPs. For the elucidation of the distinct physiological roles of the β2c and β2i subunits site-specific probes or subunit-selective compounds would be required. So far, mice lacking subunit β2i were demonstrated to be not affected by DSS-induced colitis [2] and drugs targeting both TL active sites were proven to sensitize malignant cells for inhibition of the β5 active sites of the proteasome [3]. Together, these preliminary results suggest a therapeutically relevant function also for the β2 subunits, which both exert rather broad substrate specificities.

According to previous suggestions [4], the unprimed substrate binding sites of subunit β1i are lined with apolar protein side chains that give rise to a Braap activity [5]. Thus, incorporation of subunit β1i into the iCP stimulates the generation of high-affinity MHC I peptides. In agreement, the cytotoxic T cell responses of β1i-deficient mice are altered with respect to their antigen specificity [6].

The exchange of subunit β5c by β5i is suggested to generally enhance peptide bond cleavage due to the unique hydrophilicity in proximity to the active site Thr1 of subunit β5i [7]. In particular, the strictly conserved amino acids Ser46

E. M. Huber, *Structural and Functional Characterization of the Immunoproteasome*, Springer Theses, DOI: 10.1007/978-3-319-01556-9_5,

and Thr127 are suggested to stimulate protein degradation. In agreement the 26S immunoproteasome was attributed twice the activity of the constitutive 26S proteasome [8] and the rates of antigen processing were shown to influence the immunogenicity of epitopes [9]. Despite similar primary sequences for the β5c and β5i substrate specificity pockets both subunits still differ in their cleavage patterns [10]. Distinct conformations of Met45 cause subunit β5c to preferentially bind tiny non-polar amino acids such as alanine or valine in the S1 pocket, while subunit β5i favours bulky aromatic P1 residues like phenylalanine, tyrosine [11] and tryptophan. These substrate preferences are in line with the published selective fluorogenic AMC-substrates Ac-Trp-Leu-Ala-AMC for subunit β5c and Ac-Ala-Asn-Trp-AMC for subunit β5i [12]. Leucine and isoleucine are supposed to target subunit β5c with a slight preference over β5i. Both presumably displace Met45 in subunit β5c, but might also not be sufficiently stabilized in the spacious S1 pocket of subunit β5i during initial ligand docking. By contrast, Tyr fits well in the S1 pocket of subunit β5i and is exceptionally well-stabilized in the β5c counterpart by interactions with Ser53 (β5c) and Ser129 (β6). Therefore, it is accepted by both β5c and β5i as a P1 residue. In conclusion, subunit β5i evolved to efficiently support antigen presentation by producing epitopes with a broad range of hydrophobic C-terminal amino acids for tight binding to MHC I receptors. Its outstanding role for antigen generation is corroborated by a 50 % reduction in MHC I levels in β5i-deficient mice [13] and their enhanced predisposition to infections [14]. Notably, the deletion of the subunits β1i or β2i does not affect MHC I expression [15–17].

Mutagenesis experiments aimed at mimicking the key features of subunit β5i in yeast in order to investigate in more detail the subtle differences between the β5 active sites. The created mutants addressed all sequence differences in the substrate binding pockets and around the catalytic Thr1. However, none of the mutant CPs was as susceptible to inhibition by ONX 0914 as subunit β5i and in agreement their X-ray structures depicted no change in the side chain conformation of Met45. These findings point out that indeed the structural differences between the β5c and β5i subunits provoke their different substrate specificities. Furthermore, the structural and functional key features of the iCP are based rather on long-range effects and amino acid networks than on single point mutations. Remarkably, substitution of the yβ5 subunit by the mammalian β5i entity is lethal to yeast even when using codon-optimized sequences and even when the yβ5 propeptide replaces the β5i counterpart (unpublished data of W. Heinemeyer). Analysis of the subunit contacts in the yCP and iCP suggest that solely Arg57 might suppress assembly of subunit β5i into the yCP by clashing with the adjacent Arg82 from subunit yβ6, but even a β5i R57T mutant could not be successfully incorporated into the yCP (unpublished results of W. Heinemeyer).

Unlike the β5i subunit, the β5c active site presumably has been successfully imitated in yeast. Mutation of Gln53 to Ser enhanced the selectivity for leucine residues in P1 without any obvious structural changes. Thus, amino acid 53 plays a crucial role in determining the cleavage preference of β5c and β5i subunits by modulating the interaction strength with Met45. Notably, this observation does probably not apply to subunit β5t, as Thr cannot adopt diverse side chain conformations like Met45 and does not interact with Ala53, which is strictly conserved among β5t sequences.

Importantly, for all CP inhibition experiments the fluorogenic substrate Suc-Leu-Leu-Val-Tyr-AMC was used, irrespective of the mutations introduced in subunit yβ5. For a more correct determination of IC_{50} values specific fluorogenic peptide substrates for the subunits β5c, β5i and β5t would be appropriate.

5.2 Guidelines for the Rational Design of CP-selective Inhibitors

Various diseases, including autoimmune disorders and cancers, are characterized by elevated levels of inflammation signals and iCP subunits [18]. Hence, iCP-selective compounds constitute promising novel drugs for the treatment of autoimmune diseases. The iCP-selective inhibitor ONX 0914 was proven to halt disease progression in rheumatoid arthritis, experimental colitis, systemic lupus erythematosus and Hashimoto's thyroiditis by decreasing the levels of proinflammatory cytokines as well as autoantibodies and by modulating cytotoxic T cell responses [2, 19–21].

Previous studies demonstrated that peptide based inhibitors of the CP including ONX 0914 mimic the binding mode of natural substrates by forming an antiparallel β sheet in the substrate binding channels of the active sites [4]. The herein presented crystal structures now elucidated that the binding mechanism of ONX 0914 to all active sites of the proteasome, the c and i subunits, is identical. Furthermore the selectivity of compounds solely depends on the potency of their electrophilic head group and the interactions of the ligand's P sites with the surrounding protein side chains. The exceptional importance of the interaction between the P1 site and the S1 pocket revealed by this study is corroborated by the nonpeptidic proteasome inhibitors salinosporamide A (marizomib) and omuralide that both occupy only the S1 site [4, 22]. Bearing in mind the uniform proteolytic mechanism of all active proteasome subunits, subunit-specific inhibitory compounds can only be developed by varying the side chains of peptidomimetics: β1c targeting compounds require acidic amino acids in their P1 site, while the hydrophobicity of the S1 pocket in β1i leads to a preference for branched hydrophobic residues, such as Val, Ile or Leu. Moreover, subunit β1i favours smaller and more polar amino acids in the S3 pocket than subunit β1c. Subunit β5c is perfectly suited to accommodate small apolar residues in P1, while subunit β5i is adapted for binding aromatic amino acids. The P3 site of subunit β5c accepts bulky hydrophobic side chains, whereas the P3 site of subunit β5i demands for small and preferentially polar residues. Since the S2 sites of the substrate binding channels are either shallow (subunit β1c/i, β2c/i and β5i) or even lacking (β5c), the P2 positions can be occupied by spacious side chains of diverse chemical nature.

Ligand complex structures revealed that binding of the phenyl P1 side chain of ONX 0914 to subunit β5c is primarily disfavoured by sterical hindrance with Met45. Consistently, non-natural derivatives of salinosporamide A and omuralide, bearing a phenyl moiety in the P1 site, demonstrate less affinity for the subunits yβ5 and β5c compared to their natural counterparts [23, 24]. Besides ONX 0914, the epoxyketone PR-924 (Fig. 1.7) represents another β5i-selective compound with phenylalanine as P1 side chain and alanine as P3 residue. However, unlike ONX 0914 and PR-924, the epoxyketone oprozomib (ONX 0912; (Fig. 1.6) that also bears a phenyl group as P1 residue is a potent inhibitor of both β5c and β5i with even slight preferences for the c subunit. The iCP crystal structure provides a plausible explanation for this observation. The methoxyserine of ONX 0912 fits into the S3 pocket of the β5c subunit formed by Ala27, but Ser27 in subunit β5i diminishes the size of the β5i S3 pocket and thus, electrostatically and sterically hinders binding of ONX 0912 to β5i. In this regard, even though the P1 residue appears to be the major determinant for the affinity of a ligand, the P2 and P3 residues also significantly contribute to the IC_{50} values. PR-825 (Fig. 1.7) a structural analogue of ONX 0912 that possesses a leucine instead of a phenyl residue in P1, targets subunit β5c with 20-fold selectivity over β5i, demonstrating that leucine and methoxyserine are suited to preferentially bind to the cCP. Carfilzomib, carrying a leucine side chain in both P1 and P3 sites is reported to be 5-times more selective for β5c than β5i [25]. Hence, leucine in P1 and leucine or methoxyserine in P3 give rise to β5c selectivity. Moreover, the yCP:bortezomib complex structure [26] together with the cCP and iCP coordinates elucidate that the P1 leucine and the P3 pyrazine ring of bortezomib fit into subunit β5i and β5c by interacting with Thr21, Ala22 and Ala27/Ser27 of β5c/i and Asp144 of the neighbouring subunit β6 (Fig 5.1). Hereby, bortezomib is capable of potently inhibiting both the cCP and iCP.

Bearing in mind a sequence identity of more than 90 % between murine and human subunits, the design of novel selective proteasome inhibitors for single CP-types and CP-subunits is now amenable. Such compounds enable the detailed examination of the biological impact of each active proteasome subunit and have huge medicinal potentials for the treatment of cancer and autoimmune diseases.

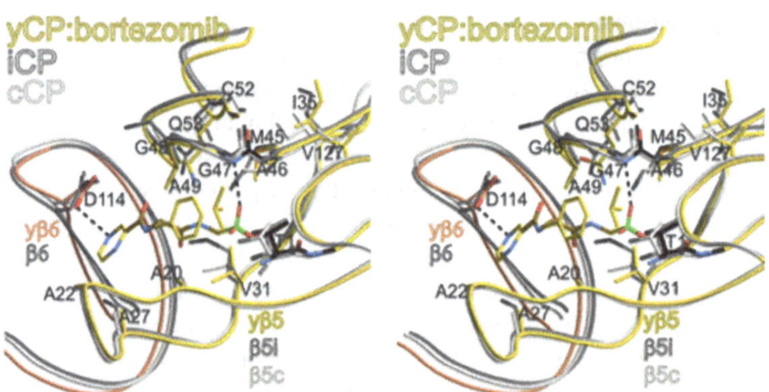

Fig. 5.1 Stereo illustration of the yCP:bortezomib complex structure superimposed onto the cCP and iCP. Structural comparison of the bortezomib bound yβ5 subunit and the unliganded murine β5c and β5i active sites. Amino acids are labelled for the yβ5 subunit. Hydrogen bonds between bortezomib and the surrounding protein residues are indicated by black dashed lines. Adapted from Huber et al., 2012 [7]

References

1. H.G. Rammensee, T. Friede, S. Stevanovi, MHC ligands and peptide motifs: first listing. Immunogenetics **41**, 178–228 (1995)
2. M. Basler, M. Dajee, C. Moll, M. Groettrup, C.J. Kirk, Prevention of experimental colitis by a selective inhibitor of the immunoproteasome. J. Immunol. **185**, 634–641 (2010)
3. A.C. Mirabella, A.A. Pletnev, S.L. Downey, B.I. Florea, T.B. Shabaneh, M. Britton, M. Verdoes, D.V. Filippov, H.S. Overkleeft, A.F. Kisselev, Specific cell-permeable inhibitor of proteasome trypsin-like sites selectively sensitizes myeloma cells to bortezomib and carfilzomib. Chem. Biol. **18**, 608–618 (2011)
4. M. Groll, L. Ditzel, J. Löwe, D. Stock, M. Bochtler, H.D. Bartunik, R. Huber, Structure of 20S proteasome from yeast at 2.4 Å resolution. Nature **386**, 463–471 (1997)
5. M. Orlowski, C. Cardozo, C. Michaud (1993). Evidence for the presence of five distinct proteolytic components in the pituitary multicatalytic proteinase complex. Properties of two components cleaving bonds on the carboxyl side of branched chain and small neutral amino acids, Biochemistry, 32, 1563–1572
6. W. Chen, C.C. Norbury, Y. Cho, J.W. Yewdell, J.R. Bennink, Immunoproteasomes shape immunodominance hierarchies of antiviral CD8(+) T cells at the levels of T cell repertoire and presentation of viral antigens. J. Exp. Med. **193**, 1319–1326 (2001)
7. E. Huber, M. Basler, R. Schwab, W. Heinemeyer, C.J. Kirk, M. Groettrup, M. Groll, Immuno- and constitutive proteasome crystal structures reveal differences in substrate and inhibitor specificity. Cell **148**, 727–738 (2012)
8. U. Seifert, L.P. Bialy, F. Ebstein, D. Bech-Otschir, A. Voigt, F. Schroter, T. Prozorovski, N. Lange, J. Steffen, M. Rieger, U. Kuckelkorn, O. Aktas, P.M. Kloetzel, E. Kruger, Immunoproteasomes preserve protein homeostasis upon interferon-induced oxidative stress. Cell **142**, 613–624 (2010)
9. P. Deol, D.M. Zaiss, J.J. Monaco, A.J. Sijts, Rates of processing determine the immunogenicity of immunoproteasome-generated epitopes. J. Immunol. **178**, 7557–7562 (2007)
10. R.E. Toes, A.K. Nussbaum, S. Degermann, M. Schirle, N.P. Emmerich, M. Kraft, C. Laplace, A. Zwinderman, T.P. Dick, J. Muller, B. Schonfisch, C. Schmid, H.J. Fehling, S. Stevanovic, H.G. Rammensee, H. Schild, Discrete cleavage motifs of constitutive and immunoproteasomes revealed by quantitative analysis of cleavage products. J. Exp. Med. **194**, 1–12 (2001)
11. M. Gaczynska, K.L. Rock, T. Spies, A.L. Goldberg, Peptidase activities of proteasomes are differentially regulated by the major histocompatibility complex-encoded genes for LMP2 and LMP7. Proc. Natl. Acad. Sci. U S A **91**, 9213–9217 (1994)
12. C. Blackburn, K.M. Gigstad, P. Hales, K. Garcia, M. Jones, F.J. Bruzzese, C. Barrett, J.X. Liu, T.A. Soucy, D.S. Sappal, N. Bump, E.J. Olhava, P. Fleming, L.R. Dick, C. Tsu, M.D. Sintchak, J.L. Blank, Characterization of a new series of non-covalent proteasome inhibitors with exquisite potency and selectivity for the 20S beta5-subunit. Biochem. J. **430**, 461–476 (2010)
13. H.J. Fehling, W. Swat, C. Laplace, R. Kuhn, K. Rajewsky, U. Muller, H. von Boehmer, MHC class I expression in mice lacking the proteasome subunit LMP-7. Science **265**, 1234–1237 (1994)
14. L. Tu, C. Moriya, T. Imai, H. Ishida, K. Tetsutani, X. Duan, S. Murata, K. Tanaka, C. Shimokawa, H. Hisaeda, K. Himeno, Critical role for the immunoproteasome subunit LMP7 in the resistance of mice to Toxoplasma gondii infection. Eur. J. Immunol. **39**, 3385–3394 (2009)
15. M. Groettrup, C.J. Kirk, M. Basler, Proteasomes in immune cells: more than peptide producers? Nat. Rev. Immunol. **10**, 73–78 (2010)
16. L. Van Kaer, P.G. Ashton-Rickardt, M. Eichelberger, M. Gaczynska, K. Nagashima, K.L. Rock, A.L. Goldberg, P.C. Doherty, S. Tonegawa, Altered peptidase and viral-specific T cell response in LMP2 mutant mice. Immunity **1**, 533–541 (1994)
17. M. Basler, J. Moebius, L. Elenich, M. Groettrup, J.J. Monaco, An altered T cell repertoire in MECL-1-deficient mice. J. Immunol. **176**, 6665–6672 (2006)
18. E.M. Huber, M. Groll, Inhibitors for the immuno- and constitutive proteasome: current and future trends in drug development. Angew. Chem. Int. Ed. Engl. **51**, 8708–8720 (2012)

19. T. Muchamuel, M. Basler, M.A. Aujay, E. Suzuki, K.W. Kalim, C. Lauer, C. Sylvain, E.R. Ring, J. Shields, J. Jiang, P. Shwonek, F. Parlati, S.D. Demo, M.K. Bennett, C.J. Kirk, M. Groettrup, A selective inhibitor of the immunoproteasome subunit LMP7 blocks cytokine production and attenuates progression of experimental arthritis. Nat. Med. **15**, 781–787 (2009)
20. H.T. Ichikawa, T. Conley, T. Muchamuel, J. Jiang, S. Lee, T. Owen, J. Barnard, S. Nevarez, B.I. Goldman, C.J. Kirk, R.J. Looney, J.H. Anolik, Novel proteasome inhibitors have a beneficial effect in murine lupus via the dual inhibition of type i interferon and autoantibody secreting cells. Arthritis Rheum. **64**, 493–503 (2011)
21. Y. Nagayama, M. Nakahara, M. Shimamura, I. Horie, K. Arima, N. Abiru, Prophylactic and therapeutic efficacies of a selective inhibitor of the immunoproteasome for Hashimoto's thyroiditis, but not for graves hyperthyroidism, in mice. Clin. Exp. Immunol., **168**, 268–273 (2012)
22. M. Groll, R. Huber, B.C. Potts, Crystal structures of Salinosporamide A (NPI-0052) and B (NPI-0047) in complex with the 20S proteasome reveal important consequences of beta-lactone ring opening and a mechanism for irreversible binding. J. Am. Chem. Soc. **128**, 5136–5141 (2006)
23. E.J. Corey, W.D. Li, Total synthesis and biological activity of lactacystin, omuralide and analogs. Chem. Pharm. Bull. **47**, 1–10 (1999)
24. M. Nett, T.A. Gulder, A.J. Kale, C.C. Hughes, B.S. Moore, Function-oriented biosynthesis of beta-lactone proteasome inhibitors in Salinispora tropica. J. Med. Chem. **52**, 6163–6167 (2009)
25. S.D. Demo, C.J. Kirk, M.A. Aujay, T.J. Buchholz, M. Dajee, M.N. Ho, J. Jiang, G.J. Laidig, E.R. Lewis, F. Parlati, K.D. Shenk, M.S. Smyth, C.M. Sun, M.K. Vallone, T.M. Woo, C.J. Molineaux, M.K. Bennett, Antitumor activity of PR-171, a novel irreversible inhibitor of the proteasome. Cancer Res. **67**, 6383–6391 (2007)
26. M. Groll, C.R. Berkers, H.L. Ploegh, H. Ovaa, Crystal structure of the boronic acid-based proteasome inhibitor bortezomib in complex with the yeast 20S proteasome. Structure **14**, 451–456 (2006)

Appendix

Table A.1 PISA analysis of interface areas and interactions of adjacent proteasome subunits in the murine cCP and iCP

Subunit		iCP			cCP			iCP compared to cCP		
1	2	Interface [Å²]	hb	sb	Interface [Å²]	hb	sb	Interface [%]	Δhb	Δsb
α1	α2	1399.9	18	3	1403.3	17	3	99.8	+1	0
α1	α7	1323.6	19	4	1341.2	16	3	98.7	+3	+1
α2	α3	1459.1	16	5	1462.1	18	5	99.8	−2	0
α3	α4	1430.3	20	5	1500.0	18	6	95.4	+2	−1
α4	α5	1395.6	21	4	1423.8	25	9	98.0	−4	−5
α5	α6	1327.8	19	4	1336.7	18	4	99.3	+1	0
α6	α7	1367.0	19	1	1365.4	20	2	100.1	−1	−1
β1(i)	β1(i)′	750.8	8	6	945.7	8	8	79.4	0	−2
β1(i)	β2(i)	471.6	5	7	552.8	8	5	85.3	−3	+2
β1(i)	β7	717.4	11	8	715.7	13	8	100.2	−2	0
β1(i)′	β7	1110.7	16	3	1142.4	19	2	97.2	−3	+1
β2(i)	β3	1531.0	25	12	1601.7	25	11	95.6	0	+1
β2(i)	β7′	694.4	8	11	744.0	7	7	93.3	+1	+4
β3	β4	752.1	12	4	737.6	13	5	102.0	−1	−1
β3	β5(i)′	855.7	19	8	841.6	19	8	101.7	0	0
β4	β4′	734.0	14	0	761.7	11	0	96.4	+3	0
β4	β5(i)	589.4	3	5	590.0	4	4	99.9	−1	+1
β4′	β5(i)	597.0	8	7	654.3	7	10	91.2	+1	−3
β5(i)	β6	688.7	12	4	547.2	10	4	125.9	+2	0
β6	β7	970.8	13	1	974.3	13	3	99.6	0	−2
β6′	β2(i)	1130.6	15	4	1215.8	16	5	93.0	−1	−1
β6′	β3	657.6	8	5	673.8	10	6	97.6	−2	−1
β1(i)	α1	457.4	5	3	417.0	3	0	109.7	+2	+3
β1(i)	α7	457.8	7	3	440.7	2	1	103.9	+5	+2
β2(i)	α1	566.2	8	6	583.5	7	5	97.0	+1	+1
β2(i)	α2	441.8	3	0	463.3	5	0	95.4	−2	0
β3	α2	412.6	8	6	427.7	10	6	96.5	−2	0

The number of hydrogen bonds (hb) and salt bridges (sb) as well as their differences between cCP and iCP (Δ) are provided. Interface areas of iCP subunits are given as percentage of the corresponding values in the cCP

E. M. Huber, *Structural and Functional Characterization of the Immunoproteasome*, Springer Theses, DOI: 10.1007/978-3-319-01556-9,

Table A.2 X-ray data collection and refinement statistics of structures of the yCP mutant yβ5 Q53S

	yCP	yCP	yCP
	yβ5 Q53S	yβ5 Q53S:bortezomib	yβ5 Q53S:ONX 0914
Crystal parameters			
Space group	$P2_1$	$P2_1$	$P2_1$
Cell constants	a = 134.7 Å	a = 136.3 Å	a = 136.7 Å
	b = 301.9 Å	b = 301.4 Å	b = 301.5 Å
	c = 144.7 Å	c = 145.2 Å	c = 145.2 Å
	β = 112.9°	β = 113.1°	β = 112.8°
CPs / AU[a]	1	1	1
Data collection			
Beam line	X06SA, SLS	X06SA, SLS	X06SA, SLS
Wavelength (Å)	1.0	1.0	1.0
Resolution range (Å)[b]	49–2.9 (3.0–2.9)	49–3.0 (3.1–3.0)	49–3.4 (3.5–3.4)
No. observations	717696	653661	521471
No. unique reflections[c]	230488	209442	146379
Completeness (%)[b]	98.2 (98.6)	97.5 (99.1)	98.6 (99.0)
R_{merge} (%)[b, d]	7.9 (55.3)	8.4 (62.0)	13.6 (58.7)
I/σ (I)[b]	11.8 (2.6)	13.0 (3.2)	7.6 (2.3)
Refinement (REFMAC5)			
Resolution range (Å)	15–2.9	15–3.0	15–3.4
No. refl. working set	218963	308271	176276
No. refl. test set	10948	15413	8813
No. non hydrogen	50864	51032	51116
No. of ligand atoms	–	168	294
Water molecules	1322	1322	1322
R_{work}/R_{free} (%)[e]	15.1/21.1	14.7/20.6	14.2/21.7
r.m.s.d. bond (Å)/(°)[f]	0.011/1.599	0.012/1.637	0.011/1.591
Average B-factor ($Å^2$)	61.7	64.5	70.7
Ramachandran plot (%)[g]	94.3/4.7/1.0	94.4/4.9/0.7	92.3/6.6/1.1

[a]Asymmetric unit

[b]The values in parentheses of resolution range, completeness, R_{merge} and I/σ (I) correspond to the last resolution shell

[c]Friedel pairs were treated as identical reflections

[d]$R_{merge}(I) = \Sigma_{hkl} \Sigma_j \mid [I(hkl)_j - I(hkl)] \mid \Sigma_{hkl} I_{hkl}$, where $I(hkl)_j$ is the jth measurement of the intensity of reflection hkl and <I(hkl)> is the average intensity

[e]$R = \Sigma_{hkl} \mid |F_{obs}| - |F_{calc}| \mid / \Sigma_{hkl} |F_{obs}|$, where R_{free} is calculated for a randomly chosen 5 % of reflections, which were not used for structure refinement, and R_{work} is calculated for the remaining reflections

[f]Deviations from ideal bond lengths/angles

[g]Number of residues in favoured, allowed or outlier region

Table A.3 X-ray data collection and refinement statistics of structures of the yCP mutant yβ5 A27S K32N A46S T57R G48C K71G V127T

	yCP	yCP	yCP
	yβ5 A27S K32N A46S T57R G48C K71G V127T	yβ5 A27S K32N A46S T57R G48C K71G V127T:bortezomib	yβ5 A27S K32N A46S T57R G48C K71G V127T:ONX 0914
Crystal parameters			
Space group	$P2_1$	$P2_1$	$P2_1$
Cell constants	a = 134.5 Å	a = 136.5 Å	a = 136.6 Å
	b = 301.0 Å	b = 300.4 Å	b = 300.3 Å
	c = 144.4 Å	c = 145.6 Å	c = 145.8 Å
	β = 113.0°	β = 113.1°	β = 113.2°
CPs / AU[a]	1	1	1
Data collection			
Beam line	X06SA, SLS	X06SA, SLS	X06SA, SLS
Wavelength (Å)	1.0	1.0	1.0
Resolution range (Å)[b]	49–2.5 (2.6–2.5)	49–2.8 (2.9–2.8)	49–2.6 (2.7–2.6)
No. observations	1048068	813713	1001216
No. unique reflections[c]	353449	260234	324245
Completeness (%)[b]	97.3 (98.4)	98.5 (99.0)	98.3 (99.4)
R_{merge} (%)[b, d]	7.5 (59.7)	7.5 (46.2)	7.6 (56.0)
I/σ (I)[b]	9.2 (2.0)	11.9 (3.4)	10.8 (2.7)
Refinement (REFMAC5)			
Resolution range (Å)	15–2.5	15–2.8	15–2.6
No. refl. working set	335776	247222	308031
No. refl. test set	16788	12361	15401
No. non hydrogen	50888	51047	51135
No. of ligand atoms	–	168	294
Water molecules	1340	1340	1340
R_{work}/R_{free} (%)[e]	19.0/23.6	15.5/21.0	16.9/21.5
r.m.s.d. bond (Å)/(°)[f]	0.012/1.652	0.011/1.551	0.012/1.649
Average B-factor ($Å^2$)	58.1	62.7	62.1
Ramachandran Plot (%)[g]	94.4/4.8/0.8	95.5/3.7/0.7	95.6/3.8/0.6

[a]Asymmetric unit

[b]The values in parentheses of resolution range, completeness, R_{merge} and I/σ (I) correspond to the last resolution shell

[c]Friedel pairs were treated as identical reflections

[d]$R_{merge}(I) = \Sigma_{hkl}\Sigma_j \mid [I(hkl)_j - I(hkl)] \mid \Sigma_{hkl} I_{hkl}$, where $I(hkl)_j$ is the jth measurement of the intensity of reflection hkl and <I(hkl)> is the average intensity

[e]$R = \Sigma_{hkl} \mid |F_{obs}| - |F_{calc}| \mid / \Sigma_{hkl} |F_{obs}|$, where R_{free} is calculated for a randomly chosen 5 % of reflections, which were not used for structure refinement, and R_{work} is calculated for the remaining reflections

[f]Deviations from ideal bond lengths/angles

[g]Number of residues in favoured, allowed or outlier region

Table A.4 X-ray data collection and refinement statistics of structures of the yCP mutant yβ5 M45R

	yCP	yCP	yCP
	yβ5 M45R	yβ5 M45R:bortezomib	yβ5 M45R:ONX 0914
Crystal parameters			
Space group	$P2_1$	$P2_1$	$P2_1$
Cell constants	a = 134.6 Å	a = 136.0 Å	a = 135.9 Å
	b = 302.8 Å	b = 300.3 Å	b = 298.1 Å
	c = 145.2 Å	c = 145.1 Å	c = 144.4 Å
	β = 112.7°	β = 112.8°	β = 112.7°
CPs/AU[a]	1	1	1
Data collection			
Beam line	X06SA, SLS	X06SA, SLS	X06SA, SLS
Wavelength (Å)	1.0	1.0	1.0
Resolution range (Å)[b]	49–3.0 (3.1–3.0)	49–2.9 (3.0–2.9)	48–2.8 (2.9–2.8)
No. observations	640110	700484	747601
No. unique reflections[c]	208693	230999	248239
Completeness (%)[b]	97.7 (98.9)	97.7 (98.6)	95.6 (96.1)
R_{merge} (%)[b, d]	10.6 (49.4)	8.7 (55.5)	7.4 (50.6)
I/σ (I)[b]	8.7 (2.7)	10.1 (1.7)	11.1 (1.8)
Refinement (REFMAC5)			
Resolution range (Å)	15–3.0	15–2.9	15–2.8
No. refl. working set	198258	219448	235827
No. refl. test set	9912	10972	11791
No. non hydrogen	50894	51053	51137
No. of ligand atoms (ligand; MES)	–	168	318
Water molecules	1340	1340	1340
R_{work}/R_{free} (%)[e]	14.8/20.7	15.9/21.7	17.5/23.9
r.m.s.d. bond (Å)/(°)[f]	0.014/1.802	0.012/1.658	0.014/1.830
Average B-factor ($Å^2$)	59.2	61.0	64.1
Ramachandran Plot (%)[g]	94.2/5.1/0.8	95.1/4.2/0.8	93.7/5.4/0.8

[a] Asymmetric unit

[b] The values in parentheses of resolution range, completeness, R_{merge} and I/σ (I) correspond to the last resolution shell

[c] Friedel pairs were treated as identical reflections

[d] $R_{merge}(I) = \Sigma_{hkl}\Sigma_j \mid [I(hkl)_j - I(hkl)] \mid \Sigma_{hkl} I_{hkl}$, where $I(hkl)_j$ is the jth measurement of the intensity of reflection hkl and $\langle I(hkl) \rangle$ is the average intensity

[e] $R = \Sigma_{hkl} \mid |F_{obs}| - |F_{calc}| \mid / \Sigma_{hkl} |F_{obs}|$, where R_{free} is calculated for a randomly chosen 5 % of reflections, which were not used for structure refinement, and R_{work} is calculated for the remaining reflections

[f] Deviations from ideal bond lengths/angles

[g] Number of residues in favoured, allowed or outlier region

Table A.5 X-ray data collection and refinement statistics of structures of the yCP mutant yβ5 I35T M45R

	yCP	yCP	yCP
	yβ5 I35T M45R	yβ5 I35T M45R:bortezomib	yβ5 I35T M45R:ONX 0914
Crystal parameters			
Space group	$P2_1$	$P2_1$	$P2_1$
Cell constants	a = 133.9 Å	a = 135.1 Å	a = 135.5 Å
	b = 300.6 Å	b=301.0 Å	b = 299.1 Å
	c = 144.2 Å	c = 146.1 Å	c = 145.7 Å
	β = 112.8°	β = 112.8°	β = 112.9°
CPs/AU[a]	1	1	1
Data collection			
Beam line	X06SA, SLS	X06SA, SLS	X06SA, SLS
Wavelength (Å)	1.0	1.0	1.0
Resolution range (Å)[b]	49–3.0 (3.1–3.0)	50–2.9 (3.0–2.9)	50–3.1 (3.2–3.1)
No. observations	621044	698969	578722
No. unique reflections[c]	205091	231926	186814
Completeness (%)[b]	97.9 (98.3)	97.7 (99.1)	96.8 (98.7)
R_{merge} (%)[b, d]	11.9 (52.2)	9.6 (50.3)	9.8 (46.8)
I/σ (I)[b]	8.6 (2.9)	8.6 (2.6)	9.8 (2.5)
Refinement (REFMAC5)			
Resolution range (Å)	15–3.0	15–2.9	15–3.1
No. refl. working set	194836	220329	177472
No. refl. test set	9741	11016	8873
No. non hydrogen	50891	51051	51161
No. of ligand atoms (ligand; MES)	–	168	318
Water molecules	1340	1340	1340
R_{work}/R_{free} (%)[e]	15.6/22.4	15.6/21.8	14.3/21.1
r.m.s.d. bond (Å)/(°)[f]	0.011/1.592	0.012/1.664	0.012/1.754
Average B-factor ($Å^2$)	55.7	62.8	61.3
Ramachandran Plot (%)[g]	94.7/4.4/0.9	94.5/4.6/0.9	93.7/5.3/1.0

[a]Asymmetric unit

[b]The values in parentheses of resolution range, completeness, R_{merge} and I/σ (I) correspond to the last resolution shell

[c]Friedel pairs were treated as identical reflections

[d]$R_{merge}(I) = \Sigma_{hkl} \Sigma_j \mid [I(hkl)_j - I(hkl)] \mid \Sigma_{hkl} I_{hkl}$, where $I(hkl)_j$ is the jth measurement of the intensity of reflection hkl and <I(hkl)> is the average intensity

[e]$R = \Sigma_{hkl} \mid |F_{obs}| - |F_{calc}| \mid / \Sigma_{hkl} |F_{obs}|$, where R_{free} is calculated for a randomly chosen 5 % of reflections, which were not used for structure refinement, and R_{work} is calculated for the remaining reflections

[f]Deviations from ideal bond lengths/angles

[g]Number of residues in favoured, allowed or outlier region

Table A.6 X-ray data collection and refinement statistics of structures of the β5t-mimicking yCP mutant

	yCP	yCP
	yβ5 A20S, A22C, V31S, M45T, A46S, G48T	yβ5 A20S, A22C, V31S, M45T, A46S, G48T:ONX 0914
Crystal parameters		
Space group	$P2_1$	$P2_1$
Cell constants	a = 135.0 Å	a = 136.6 Å
	b = 302.5 Å	b = 299.9 Å
	c = 144.2 Å	c = 146.1 Å
	β = 112.8°	β = 113.2°
CPs/AU[a]	1	1
Data collection		
Beam line	X06SA, SLS	X06SA, SLS
Wavelength (Å)	1.0	1.0
Resolution range (Å)[b]	25–2.9 (3.0–2.9)	25–3.1 (3.2–3.1)
No. observations	724106	558889
No. unique reflections[c]	230082	185554
Completeness (%)[b]	99.4 (97.9)	95.0 (97.4)
R_{merge} (%)[b, d]	11.9 (53.3)	14.3 (54.0)
I/σ (I)[b]	7.5 (2.1)	6.3 (2.2)
Refinement (REFMAC5)		
Resolution range (Å)	15–2.9	15–3.1
No. refl. working set	218577	176276
No. refl. test set	10928	8813
No. non hydrogen	50876	51129
No. of ligand atoms	0	294
Water molecules	1322	1322
R_{work}/R_{free} (%)[e]	15.4/21.2	14.7/21.7
r.m.s.d. bond (Å)/(°)[f]	0.011/1.589	0.012/1.681
Average B-factor ($Å^2$)	54.9	66.5
Ramachandran Plot (%)[g]	95.7/3.7/0.6	92.9/6.0/1.1

[a]Asymmetric unit

[b]The values in parentheses of resolution range, completeness, R_{merge} and I/σ (I) correspond to the last resolution shell

[c]Friedel pairs were treated as identical reflections

[d]$R_{merge}(I) = \Sigma_{hkl} \Sigma_j \mid [I(hkl)_j - I(hkl)] \mid \Sigma_{hkl} I_{hkl}$, where $I(hkl)_j$ is the jth measurement of the intensity of reflection hkl and <I(hkl)> is the average intensity

[e]$R = \Sigma_{hkl} \mid |F_{obs}| - |F_{calc}| \mid / \Sigma_{hkl} |F_{obs}|$, where R_{free} is calculated for a randomly chosen 5 % of reflections, which were not used for structure refinement, and R_{work} is calculated for the remaining reflections

[f]Deviations from ideal bond lengths/angles

[g]Number of residues in favoured, allowed or outlier region

Zeitfracht Medien GmbH
Ferdinand-Jühlke-Straße 7
99095 Erfurt, Deutschland
produktsicherheit@kolibri360.de